T0028458

HUMAN ORIGINS

A Short History

SARAH WILD

Michael O'Mara Books Limited

First published in Great Britain in 2023 by
Michael O'Mara Books Limited
9 Lion Yard
Tremadoc Road
London SW4 7NQ

A CIP catalogue record for this book is available
from the British Library.

Papers used by Michael O'Mara Books Limited are natural,
recyclable products made from wood grown in sustainable forests.
The manufacturing processes conform to the environmental
regulations of the country of origin.

ISBN: 978-1-78929-578-8 in hardback print format
ISBN: 978-1-78929-579-5 in ebook format

1 2 3 4 5 6 7 8 9 10

Designed and typeset by Luana Gobbo
Illustrations by Peter Liddiard
Printed and bound by CPI Group (UK) Ltd, Croydon, CR0 4YY

www.mombooks.com

CONTENTS

For the thousands of researchers and technicians who make palaeoscience possible. Discoveries are a matter of luck. It's the people in the background who turn them into knowledge.

INTRODUCTION

In 2011, I was killing time in the break room at the Bernard Price Institute for Palaeontological Research at Wits University in Johannesburg in South Africa, waiting for an interview. I'd loitered in a number of university break rooms over the years, but this one was different: there was a giant table, covered with thousands of puzzle pieces.

It was only when I went downstairs, into the laboratories where technicians were painstakingly extracting fossils from the hard rock that surrounded them, that I understood the significance of the puzzle. One researcher had a piece of bone in each hand, and was trying – and failing – to fit them together. Each fragment and shard of skeleton was a puzzle piece that, when taken together, told the story of life on our planet. By building a picture of what these creatures looked like, the palaeontologists could say other things about them – how they moved, what they most likely ate, which other plants and animals were around at the same time. In short, they were trying to paint a history of the world using bones.

But we are missing many of the pieces of our human evolution puzzle. In fact, we lack most of them, but through careful, meticulous work – excavating specimens from the rock, describing them, dating them, asking questions about the palaeo-environment and challenging assumptions – we are uncovering more and more.

The book is organized chronologically, starting with primates' split from other mammals between 80 and 90 million years ago, so that you can follow the narrative of our evolution from its early beginnings to today.

The aim of this book is to break the complex human story into bite-sized chunks, to highlight the broad trends and the major disagreements – and there are many disagreements. In all my years of covering science and technology, I have not encountered an area of research in which so little is agreed upon.

I have also tried to highlight how much our understanding of human origins has changed – and continues to change – and how, in some ways, palaeoanthropology showcases some of the best features of scientific inquiry. New evidence displaces old theories, assertions have to be strongly defended in the face of robust criticism, and methodical, careful science triumphs over outdated biases.

We have come a long way from the blind dogmatism of the nineteenth century, in which it took decades to shift outmoded and unscientific ideas. For a field in which the primary evidence has been entombed in rock thousands and even millions of years ago, today things change surprisingly quickly in palaeoanthropology. Even during the course of writing this book, new information – such as the discovery of tools at Nyayanga in Kenya, the evidence for a pan-African origin for Sapiens, and signs that *Homo naledi* may have buried its dead and etched on the walls of the Rising Star Cave in South Africa – reshuffled parts of the human story. In the last three decades, we have discovered several new hominin species, and with each significant find the narrative of human origins has become more complex and nuanced.

The discovery and sequencing of Denisovan DNA, followed by the proof of human interbreeding, astonished most, if not all, palaeoscientists. And yet in the face of such evidence, startling though it was, the scientific consensus shifted and the human story changed. Similarly, it seems increasingly likely that our species, Sapiens, originated in multiple places in Africa, rather than a single birthplace. Throughout this book, I refer to Homo sapiens as Sapiens.

There are more scientists investigating the matter of our origins than there ever were, and they come from a greater diversity of backgrounds and countries, enriching the scholarship of the discipline.

Also, the technological advances that are taking human society by storm – such as big data and widespread genome sequencing – are disrupting the palaeosciences. Novel tools and methods, often appropriated from other disciplines, allow us to explore new fossil and artefact discoveries with ever greater precision and also re-examine old finds.

In many ways, the science and study of palaeoanthropology has developed right alongside us, and is perhaps one of the most human activities we have. I hope that this book will help you to recognize, as I have, just how awe-inspiring humans really are, and also appreciate our ingenuity and innovation as a species – both in terms of our evolution and development, and the work that has gone into understanding how we became the people we are today.

PART 1

A LONG
EVOLUTIONARY ROAD

1   PRIMATES

All living organisms contain DNA, a molecular blueprint that children inherit from their parents. Sometimes genetic information changes slightly as it is passed on. Often these mutations mean nothing – a tiny blip in the complex, long code of life – but over time and in successive generations, the changes can accumulate and set plants and animals on a different evolutionary path from their ancestors.

As time passes and juveniles grow into adults that have offspring of their own, a species can change permanently, developing into a different sub-species or even – over multiple generations – a new species altogether. Salamanders which fell into a cave in Slovenia many years ago lost their skin colour and sight, and their snouts got longer. Today, we know these populations as the European olm. Meanwhile, on the Galápagos Islands, finches developed differently shaped beaks, depending on the food they could access. These small, superficial changes help animals adapt to altered environments.

But over the course of millennia, evolutionary modifications can transform one species into another.

Human evolution studies tend to focus on the evolutionary fork in the road, between 5 and 9 million years ago, when humans and chimps parted ways. But in reality, the slow accumulation of changes had been taking place long before then. Many key genetic events had been set in motion millennia prior to our hominin ancestors walking upright.

The traits that make humans, well, human – our short thumbs adapted to grasping, hip sockets that sit firmly in our pelvis and allow us to put one foot in front of the other, and giant brains capable of introspection and music and art – had been millions of years in the making.

Based on molecular dating, scientists estimate that the primate lineage branched off from other mammals between 80 and 90 million years ago. One possibility for the oldest known almost-primate is the *Purgatorius*. This sharp-toothed creature was scampering through the Purgatory Hill region of Montana in the United States more than 63 million years ago. About the size of a rat, the tree-climbing mammal, known as a 'proto-primate', appeared soon after the dinosaurs became extinct. It laid the basis for Plesiadapiformes, a group of recognized early primates. Plesiadapiformes had long fingers suited to tree-climbing, and back teeth that could chew.

A contender for the title of the oldest known actual primate is *Altiatlasius koulchii*, an extinct creature that lived and died 57 million years ago in what is now Morocco in north-western Africa. It is difficult to say more about *A. koulchii* since everything we know about it is based on ten upper and lower molars and a fragment of a jawbone.

Charles Darwin: the 'tree of life' vs 'impenetrable thicket'

In 1858, English scientists Charles Darwin and Alfred Russel Wallace announced a paradigm-changing theory: creatures that were better adapted to their environment were more likely to survive and have children of their own. Called natural selection, this is the primary tool of evolution, as successive generations would selectively breed to have traits that allowed them to survive. The following year, Darwin published his highly controversial book, *On the Origin of Species*. He proposed that all species descended from a common ancestor and described all life as belonging to a single tree with multiple evolutionary branches.

While the idea of evolution is now accepted by scientists, modern genetics have called the 'tree of life' analogy into question. Rather than a tree with discrete, tidy branches, evolution is more of an impenetrable thicket, they say, with creatures cross-breeding more frequently than previously thought.

Today, there are more than five hundred different types of living primates, and new species are recognized each decade. Primate comes from the Latin for 'prime' or 'first rank'. In 1758, Swedish taxonomist Carl Linnaeus first classified the natural world into units. He grouped humans, apes and monkeys together based on their physical similarities, and labelled them the 'highest order' of mammals, hence 'primate'. Over time, and following Charles Darwin's publication of *On the Origin of Species* almost a century later, the primate category has broadened rather significantly. It is now a large and varied group of animals, ranging in size from the mouse lemur, which is as heavy as two tablespoons of sugar, through to the gorilla, which can weigh more than 200 kilograms.

There is no single characteristic that defines a primate. Rather they are categorized based on a collection of attributes. They tend to have larger brains for their body size compared to most other mammals. This cunning group is likely to rely on sight to a greater extent than smell, with forward-facing eyes that can perceive depth. (When animals, such as horses and cows, have eyes on the sides of their face, they have a wide panoramic field of vision at the expense of depth perception.)

Although primates can live in any environment, they are well adapted for life in the trees. They tend to have mobile shoulders and grasping hands and feet that allow them to swing from branch to branch.

Primates are also placental mammals, which give birth to live young, and all male primates have pendulous penises and testes that hang outside of their body. Several primate groups have upright torsos, allowing them to sit and stand with straight backs. Some primates also have tails although, notably, not apes.

Enter the anthropoids

Anthropoids, which include monkeys, apes and humans, are a special subset of primates. The group is also sometimes referred to as simians or higher primates.

Questions remain about exactly when and where this split happened – and which creature best represents this divide. Just west of the Nile River in Egypt lies the Faiyum region, a treasure trove of archaeological sites and ancient primates. Some believe that the split between anthropoids and their primate relatives is contained in a single species, Aegyptopithecus zeuxis, which lived about 38 to 29.5 million years ago in the Faiyum region. It represents a mosaic of monkey- and ape-like characteristics.

Other palaeoscientists argue that the Eosimias, found in China and Myanmar and meaning 'dawn monkey', is a better candidate for a bridge to higher simians. This sharp-toothed tree-dweller could fit in the palm of a modern human's hand and was on Earth between 45 and 40 million years ago.

These creatures, which lived in Africa and Eurasia, belonged to a group called catarrhines, and were a common ancestor for humans, apes and monkeys. They are often referred to as Old World monkeys, but many scientists reject this description when talking about the creatures at the beginning of our lineage. With some justification, they say that it is confusing as Old World monkeys such as baboons and vervet monkeys exist today (known as Cercopithecidae).

The archaeological record, with its gaping omissions and scarce clues, is shrouded in mystery. But in a field known for its leaps of logic, the existence of New World monkeys, known as platyrrhines, really does stand out. Sixty-five million years ago, South America had no monkeys. It had sloths and armadillos, but no monkeys. Suddenly (in archaeological terms), there they were.

The earliest known New World monkeys – the *Branisella boliviana*, a 1-kilogram monkey found in Bolivia, and *Canaanimico*, which was double the size of *Branisella* – have been dated to about 26 million and 26.5 million years old respectively. New World monkeys are still around, and include species such as howler and spider monkeys.

And our best hypothesis for how their ancestors arrived in South America? More than 30 million years ago, Old World monkeys boarded natural floating mats of grass and earth and travelled across the Atlantic.[1] Back then, South America and Africa were closer together – about 1,600 kilometres (1,000 miles) compared to today's 2,850 kilometres (1,770 miles) – but still a daunting distance. And it appears that ancient anthropoids made the first transatlantic crossing, helping to spread monkeys around the world.

Old World monkey versus New World monkey

How do fossils form?

Not every bone will become a fossil. In fact, the fossil remains we have are an infinitesimal fraction of the organisms and species that have lived, died and disappeared on Earth. Most species existed and went extinct without leaving a trace.

Usually, when an animal dies its soft tissue rots in the open air or is eaten or scavenged. For this reason, the overwhelming majority of fossils are of marine creatures. Sometimes when creatures in the ocean die, if they are not pulled apart by the currents or eaten, they lie on the seabed and are quickly covered by sand or silt. There, they are safe from decomposing oxygen. Layers build up, compressing the skeleton under the weight of tonnes of sediment and turning it into rock. At the same time, water seeps into the rock, and minerals in the water replace the bone. This process can take any time from 10,000 to millions of years.

While fossilization is fairly uncommon in the oceans, it is even rarer on land. Many hominin fossils have been discovered in caves or deep wells, where they were protected from scavengers and the elements. These sheltered environments also increased the likelihood of them becoming fossils – and surviving for millions of years, long after the rest of their community had turned to dust.

For the most part, Old and New World monkeys are fairly similar, but there are some notable differences. New World monkeys (platyrrhines) have broad nasal septums, which is the cartilage and bone that separate the nostrils, and their nostrils point to the sides. Some of the larger ones also had prehensile tails that could grasp things, which none of our forebears had. They also had three premolar teeth to our ancestors' two. On the other hand, Old World monkeys (our progenitors, the catarrhines) have thin septums and nostrils that point down and forward. The name 'catarrhines' quite literally means 'down nose', while 'platyrrhines' comes from the ancient Greek for 'flat nose'. Catarrhines' thumbs were more noticeably different from their other fingers compared to those of platyrrhines.

2 A SHREWDNESS OF APES

Today's modern humans, apes and Old World monkeys shared a common ancestor between 30 and 25 million years ago, but soon after that (evolutionarily speaking) humans and apes, known as hominoids, split from what we know as monkeys.

Hominoid means 'man-like', and the easiest way to tell the difference between a monkey and an ape is the presence of a tail, something which all mammals (except apes) have. All that time ago, there was a slight tweak in the TBXT gene, also known as the 'tail' gene, and ancestor apes suddenly lost their tails – a change that has carried on in their descendants, including us.[2]

That said, it is surprisingly difficult to positively identify an early long-extinct ape by its lack of tail since, well, most of the rest of the skeleton tends to be missing too. Nevertheless, the lack of a tail is an important attribute of the superfamily, known as Hominoidea. Living apes include gorillas, chimpanzees, bonobos, orangutans, gibbons and, of course, humans.

Aside from the tail, the humble elbow joint divides apes from other anthropoids (namely monkeys). If you extend your arm, your elbow is able to lock – the upper arm bone (humerus) and forearm bones (ulna and radius) fit together and are able to pivot around each other. If you put your palm face up, you may be able to rotate it 360 degrees. Such flexibility is characteristic of apes. Monkeys, on the other hand, cannot straighten their arms. With our innovative elbows and the strong muscles that run up and down our arms, we can hang from trees and also do handstands, something which doesn't come naturally to a monkey.

What is a hominoid?

Today, there are two branches of Hominoid: gibbons, which are rather dismissively referred to as 'lesser apes' (family: Hylobatidae), and the much more aggrandizing 'great apes' (Hominidae), which includes orangutans, gorillas, bonobos, chimpanzees and humans. Using molecular dating, scientists estimate that the gibbon lineage diverged from 'great apes' about 17 million years ago.

The first possible ape is the *Kamoyapithecus*, which was discovered in eastern Africa in 1948. The creature, named after renowned Kenyan field palaeontologist Kamoya Kimeu, was alive between 24 and 27 million years ago. However, there is debate whether *Kamoyapithecus* is actually an ape or whether it was still a catarrhine. All that we know about the chimpanzee-sized primate is based on its jaw and teeth, so there is room for argument.

Similarly, scientists are not certain about another candidate's status as an ape, the *Proconsul*. The tailless primate was certainly roaming around Kenya's Turkana region between 23 and 17 million years ago, but it had a mixture of catarrhine and ape characteristics: while it had the curved and flexible back of a monkey, it also had an ape-like facial structure and was able to grasp with its hands. It was named after Consul, a performing captive chimpanzee in London, and its name literally means 'before Consul'. One of the major arguments in favour of *Proconsul*'s ape status is that it had a stabilized elbow joint. Individuals have been found in Kenya and Uganda.

While modern apes share many physical and behavioural characteristics, these features evolved over time, so fossil species in this evolutionary phase are often referred to as 'stem apes' but are still included in Hominoidea.

Generally speaking, apes have wide, stable torsos, which are comparatively shallow front-to-back, and they tend to sit upright. They also have fewer lumbar vertebrae in their spine than monkeys. Their shoulders are very mobile, with blades lying on their backs rather than on the side of the body like monkeys. They also have relatively large brains, and distinctive chewing teeth in their lower jaw – five raised bumps in the shape of a Y.

Out of Africa

So far in our evolutionary narrative, ancient ape fossils have come from Africa. Primates have been found in other places, but apes have been confined to Africa. This could be a function of bias, chance, or it could be part of the human story. One theory is that climate change killed off most other primates in the planet's northern regions, preventing them from evolving into something else. A 2023 study,[3] for example, describes two new primate species that lived in the Arctic 55 million years ago, when global temperatures were warmer, but these creatures were unable to adapt when the Arctic got colder. Unlike humans, they did not have the means to protect themselves from the elements.

In general, though, ape fossils are incredibly rare because apes tended to end up as food for predators, and their bones removed by scavengers, rather than preserved for millennia. Some scientists argue that monkeys and apes were more likely to live in forest environments, which were teeming with hungry life and which would have consumed their bodies. Others say that the acidic soil environment made it unlikely for remains to fossilize. It is also possible that there are ancient ape fossils waiting to be found in other parts of the world, biding their time to write their part of Earth's history.

However, scientists do know that about 23 million years ago, an ocean separated the northern reaches of what is now Africa from Europe and Asia, and that these regions were (very slowly) drifting towards each other. There is some evidence that 20 million years ago there was a land bridge between the land masses. And 17 million years ago, when

Africa was connected to Europe, ape fossils began to appear in Germany and Turkey.

The discovery of ape fossils in the Sivalik (or Siwalik) Hills in Asia in the early twentieth century gave palaeo anthropologists new and bemusing data points to anchor the narrative of apes' migration around the world. The range lies in the foothills of the Himalayas, traversing India, Pakistan and Nepal, and the oldest specimens have been dated to about 12 million years ago. The genus found there, Sivapithecus, is thought to be an extinct ancestor of modern orangutans. Scientists think that orangutans split from African apes about 10 million years ago.

Tick-tock of the molecular clock

Modern molecular techniques continue to poke holes in our narrative of human evolution, making us rethink our origins. Back in the nineteenth and early twentieth centuries, scientists drew conclusions based on animals' morphology. But today genetic analysis can whisper secrets that bones can't.

All living things contain DNA (deoxyribonucleic acid), which is a molecule that holds genetic information. In fact, most cells in complex life forms include a full set of DNA. This information allows an organism to grow and function. Long strands of DNA, called chromosomes, comprise an organism's genome, which is effectively a full set of instructions to develop, survive and reproduce. When organisms reproduce, they pass their DNA on to their children.

The human genome contains twenty-three pairs of chromosomes, encapsulated in each cell's nucleus. We get half of our genetic information from each parent. There are also two special chromosomes which encode for our sex: X and Y chromosomes. DNA from the Y-chromosome (which is the male-specific chromosome) is passed down from father to son, and can be used to trace lineages through the male line.

Also found within animal and plant cells is a small structure called the mitochondrion, which is in charge of the cell's energy production, among other things. It has special DNA of its own, too, known as mitochondrial DNA (mtDNA). MtDNA is passed directly from a mother to her children, without any intentional manipulation or additions along the way.

Two important genetic events take place when parents have a baby: mutation and recombination. The parental DNA can mutate when it is passed on to their child and, for example, a molecule in a DNA sequence called a nucleotide can alter into a slightly different molecule. Recombination also occurs, which is when the parental DNA breaks apart and reforms again, something which is necessary when two separate sets of DNA are combined in one individual.

We know how frequently genetic mutations take place in humans in each generation – a newborn will have about seventy nucleotide changes in its genome of about 6 billion such molecules. Recombination also leads to changes in a child's DNA compared to that of their parents – about one or two per chromosome. MtDNA – surprisingly, since it is shared directly from a mother to her children – is more likely to accumulate mutations than nuclear DNA, possibly because it lacks repair mechanisms within the mitochondrion. By adding up the number of accumulated mutations, scientists can estimate how long ago two individuals – from the same or different species – shared a common ancestor.

Molecular clocks are particularly useful for tracing the movements of *Homo sapiens* around the planet, and our divergence from other recent *Homo* species. But it gets trickier the further back in time you go. Different species' molecular clocks tick at slightly different speeds (and they can even vary within populations). Also, older fathers transmit more mutations to their children than younger fathers. And those using molecular clocks have to make a number of assumptions, such as a species' child-bearing age. Scientists also think that the average mutation rate seems to have slowed over the course of human evolution.

Despite these caveats, molecular clocks are a powerful tool in palaeoanthropology, and are often used in conjunction with other dating methods to calibrate the clocks. Using mutations in mtDNA, scientists estimate that chimpanzees and hominins diverged between 5 and 9 million years ago, which is supported by the fossil record.

A tiny genetic step

Up until the mid-twentieth century, scientists only had morphology to guide their investigations, and the question of which 'great ape' was humanity's closest relative lingered in the corridors of anthropology departments.

But the proliferation and increasing sophistication of molecular technology and gene sequencing have allowed scientists to better understand our relationships with other apes. In 2005, when the chimp genome was first sequenced,[4] such methods were prohibitively expensive. Today, people can have their whole genome sequenced as part of diagnostic healthcare.

In 2023, scientists in the Zoonomia consortium sequenced the genomes of 240 mammals – from dormice and whales to bats and humans – and combed them for similarities.[5] They estimated that at least 10.7 per cent of the human genome is identical to the genomes of all the species they studied. However, we have known for a while that chimpanzees are our closest relatives, and that the extent of our genetic similarity depends on which part of our genomes you're looking at.

In terms of coding DNA – the actual genes that actually code for proteins – there is a less than 1 per cent difference between humans and chimpanzees. But large tracts of DNA do not code for anything specific (it's sometimes called 'junk DNA'), and in this DNA there's a difference of about 1.2 per cent when comparing segments of DNA that humans and chimps share. But when chimp and human genomes are considered in their entirety, there are large sequences that have been deleted and inserted, and so there is a difference of about 4 per cent.

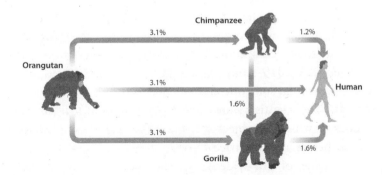

The differences in shared non-coding (aka junk) DNA in various living ape species.

For some context, while that makes us very closely related, the genetic differences between chimps and humans are about ten times that seen between any two *Homo sapiens* on the planet. So what this tells us about human evolution is that it has been a long time since humans and chimps shared a common ancestor – anywhere between 5 and 9 million years.

Meanwhile, the difference between human and gorilla non-coding (aka 'junk') DNA is 1.6 per cent and, interestingly, the difference between chimps and gorillas is also 1.6 per cent. This shows that there has been even more time to accumulate genetic mutations, indicating an even earlier common ancestor. Scientists have dated the divergence between gorillas', humans' and chimps' ancestors to about 10 million years ago. The orangutan is even further removed, with a 3.1 per cent difference in non-coding DNA. The last common ancestor of all living apes is thought to have lived about 15 million years ago.

Every year, we learn more about the intelligence and capabilities of our ape relatives. Numerous studies have shown that great apes have advanced cognitive capabilities and are, in fact, like humans in many ways. Chimps, for example, can recognize themselves in mirrors, they can use tools, communicate in complex ways and solve problems. A 2019 study even suggests that great apes can anticipate when someone else will act according to mistaken beliefs.[6] The scientists showed thirty-four great apes recorded videos in which an object was placed in one of two locked boxes, before a person tried to open the boxes.

In some of the videos, another scientist moved the object into another box, tricking the box-opener. The apes were able to put themselves into the shoes of the duped person, and anticipate which box they would open, even though the apes knew that the object had been moved. Known as a 'theory of mind', the apes understood another person's knowledge of the world (which is different to their own). Importantly, the study shows that chimpanzees can pass a similar test to the one used to assess 'theory of mind' in human children.

There are other attributes that set humans apart from our great ape relatives, such as our adaptable brains, high level of self-awareness, and advanced language and cultures, but there is no doubt that we also have a lot in common. And the more we learn about other apes, the more we recognize how similar we all are.

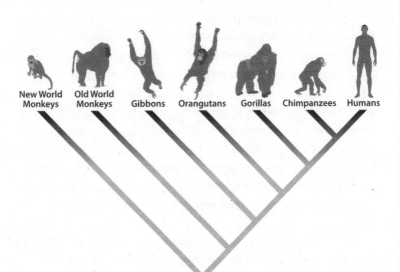

Anthropoids, which include monkeys, apes and humans, are a special subset of primates. The group is also sometimes referred to as simians or higher primates. .

Hominid versus hominin

Hominids are all living and extinct 'great apes'. They are (or were) part of the family Hominidae, which includes *Homo sapiens*, chimpanzees, gorillas, orangutans, bonobos and their extinct ancestors.

The 'African apes' group – which unsurprisingly includes ape species found in Africa – is called Homininae. But the chimp–human branch that split off, and includes living and extinct chimps (the subtribe Panina) and humans (and our ancestors), is called Hominini. The promiscuous use of the naming root 'hominin' makes shortening the names confusing, and the name 'hominin' is contested.

Some say that hominin should include chimps, gorillas and humans; others think that it should only be chimps and humans. But, in common usage, there is only one species of hominin left – us, *Homo sapiens* – and all our extinct *Australopithecus* and *Homo* ancestors are also called hominins.

When this book talks about hominins, it is referring to the lineage of species that diverged from chimpanzees and resulted in us.

Dating techniques

Dates are vital in human evolution. They anchor the narrative of our development, and a misdated fossil can completely distort the story that ancient remains tell us. With (relatively) so few hominin specimens, dates fix species, with their morphological eccentricities, to a specific place and moment in history. With this information, palaeoanthropologists are able to draw comparisons between species, and flesh out the plot of human history. At the same time, the story is just that: a narrative that we have attached to the remains to link them to each other and to us. Information such as dates provides scientific evidence for this narrative. But the remains we are dealing with are often millions of years old, and dating is one of the trickiest aspects of palaeoanthropology.

When the first hominin fossils were uncovered in the nineteenth century, we did not have the sophisticated technological toolkit we have today. Scientists mainly relied on relative dating techniques, such as geochronology, which allowed them to identify and estimate the dates of different sediment layers and sandwich the fossil's time of death approximately into those time bands. The underlying assumption is that the further down you dig, the older the layer, and so fossils entombed in deeper layers are older than those closer to the surface.

But natural disasters happen: earthquakes rip up the Earth, layers erode, and shifting tectonic plates cause the ground itself to buckle. To compensate for these confounding factors, scientists often use index fossils to define the age boundaries of different strata and show that sedimentary layers in different places are the same age. This field is known as biostratigraphy.

The basic premise of index fossils is that at a given point in time, certain organisms were abundant. These species had a wide geographic footprint and were found in a large number of places, but they also appeared and went extinct in a relatively short period. A small fraction of these creatures – plants and animals – became fossils, trapped in a stratum. Ultimately, they became signposts to indicate the age of the layer.

The flip side of this dating coin is biochronology, in which fossil species collections are not tied to a specific stratum, but indicate a given time period. In biochronology, if comparable species are found at two geographically separate sites, the locations must be about the same age. This method has been particularly useful in helping to date the hominin fossils in South Africa, where traditional dating is problematic. The animal species discovered with hominin fossils in South Africa's limestone caves, namely antelope and monkeys, are similar to those found at key eastern African sites, which are very accurately dated. Consequently, scientists were able to assign dates to the South African hominins by comparing them with the eastern African locations. A recent study looked at the tooth size of an extinct baboon and found that the size of baboon molars (specifically the *Theropithecus oswaldi* lineage) could be used to determine the geological age of South African hominin sites.[7] It should be noted, however, that this method is not particularly accurate, with estimates frequently off by a few hundred thousand years.

Palaeomagnetism is another method for dating rocks and strata, and the fossils within them. When magnetic particles lie on the Earth's surface, they fall in a specific orientation, determined by the direction and intensity of our planet's magnetic field. Over time, these magnetic minerals become trapped in the rock strata, and are held fast even when the Earth's magnetic field changes. The field has altered many times in the past, and these imprisoned magnetic minerals divulge when they were captured. Olduvai Gorge in Tanzania, once home to the famous 'Handy Man' *Homo habilis*, was the first early hominin site to be dated using magnetostratigraphy, which is when scientists use ancient planetary magnetism to constrain the age of rock strata.

Radiometric dating

With the discovery of carbon dating in the 1940s, scientists were able to give accurate dates to relatively recent fossils. Life as we know it depends on carbon – all creatures consume it in the organic matter they eat (or that their prey ate) and in the air they breathe. But not all carbon is the same. The stable carbon-12 atom (composed of six protons and six neutrons) accounts for the overwhelming majority of carbon in the atmosphere, but there is also a radioactive type of carbon, an isotope called carbon-14 (which has six protons and eight neutrons). The ratio between the two variants in the atmosphere remains stable, for the most part – and so their balance in living plants and animals is fixed. But when creatures die, they no longer consume more carbon. The radioactive carbon-14 in their bodies starts decaying and is not replaced. From the amount of carbon-14 present in a sample, scientists can estimate how long ago an organism died.

But, sadly, carbon dating has a hard limit on its usefulness. First, the specimen needs to have carbon or organic matter in it. Many artefacts in the archaeological record, such as stone, metal and many types of rock, contain no carbon. Also, over the course of about 50,000 years, carbon-14 will decay to the point that it is untraceable. In the human story, 50,000 years ago is the blink of an evolutionary eye. Modern humans had already been around for a couple of hundred thousand years. Neanderthals were living in Eurasia, the last surviving *Homo luzonensis* may have still been walking around the island of

Luzon in the Philippines, and the diminutive *Homo floresiensis* could have still lived on Flores in Indonesia. There could have been Denisovans wandering around what is now Siberia, but we're not certain. Nevertheless, all of our other hominin ancestors, from *Sahelanthropus tchadensis* 7 million years ago to each of our *Australopithecus* relatives and *Homo* cousins, had long since died. This renders carbon dating singularly useless in trying to tease apart the story of humanity's early ancestors.

Other radiometric methods, however, let palaeoscientists look further back in time. Uranium-lead (U-Pb) dating is one of the oldest means of radiometric dating and vital in palaeoanthropology. It allows scientists to determine the age of rocks that formed between 4.5 billion and 1 million years ago. Over time, uranium decays into lead, but there are two types of natural radioactive uranium atoms (with atomic masses of 235 and 238 respectively). This means that for every analysis scientists can get two ages, and then compare them. Also, scientists only need a small amount of rock – with sophisticated equipment, they can analyse samples weighing a few millionths of a gram.

Other unstable atoms also open windows into the past. With K-Ar dating, an unstable isotope of potassium (potassium-40) becomes trapped in rock as it forms, where it decays into argon-40. Newly formed rocks don't have argon in them, so by measuring the ratio of potassium to argon, scientists can work out how long ago the rock formed.

With K-Ar dating, scientists are able to date rocks from about 4.3 billion years ago (which is pretty much how long the Earth has been around) to about 100,000 years ago. Ar-Ar dating is a newer version of K-Ar dating that works on the same principle, but scientists radiate the rock sample before using a mass spectrometer to decipher its composition. In mass spectrometry, samples are converted into gases, and their constituent atoms turned into ions (an atom with an electrical charge). These ions are accelerated using an electric field, before deflecting and analysing them. The system can reveal the number of ions and their mass, allowing scientists to determine which atoms are present and how abundant they are. Such techniques are particularly well suited to identifying the age of volcanic tuffs, like those found in eastern Africa's rich hominin sites.

Uranium-series dating is also an important technique in palaeoscience's toolkit. Uranium is found in most water, so any material that forms from water (such as stalagmites in caves, or even the pigment in rock art paintings) contains uranium. Unlike K-Ar dating, in which potassium decays into the more stable argon, radioactive uranium decays into radioactive thorium, which is itself unstable. By gauging the balance between these two, scientists can work out the age of a material up to about half a million years old. With this technique, scientists can bridge the time chasm between the likes of K-Ar dating and carbon dating.

Light, electrons and the march of technology

Radioactive substances can also affect the surrounding material and point to its age. In electron spin resonance dating, for example, scientists can work out how long the mineral has been exposed to radioisotopes trapped inside it. This technique has been particularly useful in dating hominin tooth enamel.

Luminescence dating can measure the last time a given sediment was exposed to light. While not as accurate as other techniques, luminescence dating is often brought in as a dating technique of last resort or a means of calibrating other methods. For example, it has been used to try to date hominin sites in South Africa, where the breccia (rocks made up of other bits of broken minerals and stones that are cemented together) and geology make dating very tricky indeed.

Dates and dating techniques occupy a prominent position in the scientific literature about hominins, their associated sites and artefacts. Pushing the age of a hominin back a million years or forward 2 million years shuffles the narrative of human evolution and can drastically alter the trajectory of our ancestors and possible connections between species. For example, there is a great deal of debate about the origins of the first anatomically modern human, and where and when *Homo sapiens* first appeared. The new dates for Sapiens remains in Morocco have pushed the rise of modern humans back more than 100,000 years. Meanwhile, palaeoanthropologists studying eastern and South African fossils are in a continual tug of war trying to claim the emergence of early hominins, with scientific papers traded between the two camps as they use scientific evidence to back up their claims or dispute those of others.

But each year technology marches forward, and it is taking the palaeosciences along with it. In some cases, this progress means more refined and sophisticated dating techniques which increase the accuracy of the dates associated with hominin-fossil sites. But it also means that workhorse technology, such as mass spectrometers and genome sequencing, is becoming cheaper and more widespread, and more sites and artefacts can be dated with increased precision – and often re-dated. And each date provides more robust evidence for the story of humans, how we evolved and what makes us who we are.

3 THE LEAP FROM APE TO HOMININ

Today, if you were to lie chimp and human skeletons side by side, it would be easy to spot the differences. For one (important) thing, you would most likely have two complete skeletons, something palaeoanthropologists dream of and seldom have.

The physical contrasts are significant. Humans have a high, rounded and much larger brain case, along with a flat face and pointy chin, while chimps have a prominent jaw and large front teeth and canines. These variations cascade down through the skeleton. A human's foramen magnum (a passage through the base of the skull that connects the brain with the spinal cord) is balanced on top of the spine because we walk upright, and our pelvis is shorter and wider. Chimps have opposable big toes, and arms that are longer than their legs.

But between 9 and 5 million years ago, the differences

between members of the chimp and hominin lineages would have been subtle. Our early hominin ancestors would have only had a few of the attributes we recognize as characteristic of hominins (as opposed to the full suite of attributes). This makes separating early hominins from early chimpanzees quite difficult.

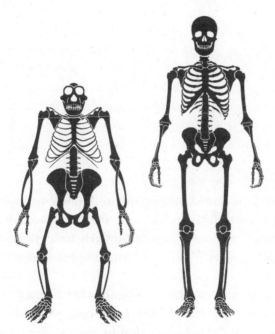

Chimpanzee versus *Homo sapiens*

Things to look for in a hominin fossil

Teeth: Living apes have very large canines and a gap in the upper jaw between the incisor and canine to accommodate the lower canines. The top canines rub along the teeth in the bottom jaw, a process called honing, which makes them sharp on the sides. Humans, on the other hand, have small canines that wear at the tips. Early hominids' teeth may have begun shrinking and changing position. Hominins also have relatively thick tooth enamel to protect their teeth from fracturing when eating hard foods.

Skull: Up until about 3 million years ago, brain size was not something that distinguished our ancestors from apes. But the location of the foramen magnum in hominins is further forward compared to chimps, because our ancestors would have spent more time upright. Something interesting to note is that, as far as we know, only *Homo sapiens* have chins. No other hominins do – and no one is sure why.

Bodies: One of the most important characteristics of hominins is that they walk upright. Known as bipedalism, walking on two feet – which non-hominins do not do for long periods of time – is something that sets humans and our ancestors apart from other apes. But the physical features that allow us to do that developed over time, and some hominin species may have been more bipedal than others. As the millennia passed, hominins evolved more vertical torsos, a shorter and wider pelvis to stabilize their footfalls, hip sockets that sit in the pelvis, straighter knees and a more stable foot.

Hands and feet: As our ancestors spent more time on two feet than in trees, our hands and feet evolved. Gorillas and chimps walk on the backs of their fingers, called knuckle walking, and have curved, strong fingers, which are ideal for climbing and holding on to branches. Humans, on the other hand, have shorter fingers with broad fingertips that allow us to manipulate tools, play the piano and hold a pencil. Hominin feet evolved to become more stable and bear the full weight of our bodies, and we lost our opposable big toe. It should be noted that if hominin fossils are rare, hominin hand and feet fossils are even rarer – they are more likely to detach from a skeleton or be removed by scavengers.

The oldest known hominin: *Sahelanthropus tchadensis*

For about 7 million years, fossilized bones and an almost intact skull lay in the Djurab Desert in northern Chad. Back then, a wooded savannah gave way to forests. By 2001, when French and Chadian palaeoanthropologists unearthed the remains of several ancient individuals, the Toros-Menalla region was blanketed with beige sand and baked under a scorching sun without a tree in sight.

Named *Sahelanthropus tchadensis*, which literally means 'the Sahel man from Chad', scientists have dated these individuals to between 6 and 7 million years old[8] – which is about the time that humanity's evolutionary branch split from that of ancient chimps. Initially, scientists estimated their age based on the other primitive animals and plants at the site (biochronology), but later narrowed down the time frame to 6.8 to 7.2 million years ago, as a result of radiometric dating.

The skull, nicknamed Toumaï (meaning 'hope of life' in the local Daza language), was significantly distorted during fossilization, but was fortunately mostly intact. Using a virtual reconstruction, scientists estimate that it would have held a brain about the same size as a chimpanzee's. That is about three times smaller than a modern human's brain. But it had other characteristics that suggest it may have been a human ancestor, rather than an ape.

Toumaï had relatively small teeth, which were worn at the tips. Human canines tend to wear at their tips, whereas other primates' canines get honed to sharp edges on the sides. By analysing Toumaï's skull, the palaeoanthropologists also found that their brainstem would have entered their skull in such a way that they could have walked upright. In apes, on the other hand, the position of the brainstem allows them to hunch and walk using their hands and feet.

However, whether *S. tchadensis* was bipedal continues to provoke debate. After the initial excavation, the shaft of a thighbone and a lower arm bone were accidentally categorized as animal bones, before being flagged in 2004 as possibly belonging to a primate. In 2017, more than fifteen years after the initial discovery, researchers began studying the thighbone to unravel its secrets. Three years later, they published a paper that revealed that the bone had most likely belonged to *S. tchadensis*, but their findings did not support the idea that the creature had walked upright.[9] A follow-up study, published in 2022, found the opposite: based on its bone density and comparisons with other living and fossil apes and hominins, the bone probably had supported (literally) upright walking – although its strong arms suggest it still spent a lot of time in trees, where it was doubtless safer from predators.[10]

Nevertheless, *S. tchadensis* offers a probable ancestor showing the jump from ape to early hominin. Then there is a 1-million-year gap in our fossil record.

Orrorin tugenensis

Thousands of kilometres east of Chad, in the Tugen Hills region of Kenya, the *Orrorin tugenensis* once roamed. 'Orrorin' means 'original man' in the Tugen language. A molar tooth was first discovered in 1974, followed by more than a dozen bone fragments almost thirty years later. The remains have been dated using K-Ar dating to between 6.2 and 5.6 million years ago, making it the second oldest hominin to have been found so far.

There is some scepticism about *Orrorin*'s inclusion in the exclusive group of human relatives. The researchers who unearthed some of the fossils have flagged its teeth as one of the reasons for its hominin status – since they have fairly thick enamel. But while the upper canine is smaller than those of apes, other researchers argue that the rest of the teeth are fairly ape-like and their enamel is not thick enough to classify *Orrorin*, for certain, as a hominin. It could, they say, be an early chimp ancestor.

But what makes palaeoanthropologists particularly excited, and why *Orrorin* is included as a hominin despite others' reservations, is the shaft of a thighbone. Its thighbone has the small ball at the top (the femoral head), a relatively flat femoral neck that links the head to the shaft of the bone, and thickened bone in the upper part of the shaft. This thick section is usually found in creatures that habitually walk upright, leading researchers to conclude that this ancient creature may have been bipedal, breaking away from the four-legged gait of apes. Bipedalism, as previously discussed, is considered a major characteristic of hominins.

Bipedalism

Apes and Japanese macaques sometimes walk upright over short distances, while lizards, some birds and, when they reach their highest speeds, cockroaches occasionally run on two legs. Kangaroos and many birds hop around on two legs. But humans alone solely walk on two legs to cover large distances. This is considered a hallmark of what it means to be human. It is certainly an ability which has allowed us to colonize the planet.

But this ability evolved over time through initially subtle changes in the spine, where it attaches to the skull and the pelvis and how the leg bones fit into the hip sockets. Hominins' legs began to adapt to walking upright – in the feet, knees and the ratio of the leg bones – while their arms and shoulders changed, as they spent more time on the ground and less in the trees. One of the major current debates is when walking became hominins' primary form of locomotion (known as obligate bipedalism) versus something they did when they had to (facultative bipedalism).

A defining characteristic of our genus *Homo* is that they had the ability to walk upright over long distances. Although not yet quite like us Sapiens, *Homo ergaster*, which lived between 1.9 and 1.5 million years ago in Africa, had the long legs and stable hips that allowed it to travel long distances and, according to mainstream thought, migrate out of Africa.

Ardipithecus

In Ethiopia, between 5.7 and 4 million years ago there lived another species of hominin, *Ardipithecus*, which means 'ground-dwelling ape' in the Afar language. Today the Awash Valley is semi-arid, but millions of years ago it would have been leafy and green, a mixture of woods and grassland, dotted with lakes and the odd swamp.

There are two proposed species of *Ardipithecus*. The oldest, *Ar. kadabba*, comes from the Middle Awash region, and all we know about it is based on some teeth, part of a lower jawbone, and fragments of a hand, foot and collarbone. Nevertheless, palaeoanthropologists have made deductions from much less.

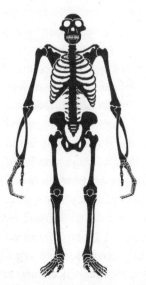

Ardi, an adult female *Ardipithecus ramidus*, lived in Ethiopia 4.4 million years ago. For many years, she was the oldest known hominin.

How to define a species: lumpers and splitters

Evolutionary anatomist Bernard Wood divides up palaeoanthropologists into two groups: lumpers and splitters. Lumpers tend to 'lump' species together, accepting that individuals within a species may look quite different, while splitters prefer to recognize many species, thus 'splitting' hominins up into numerous categories.

At the heart of this division – and the source of many arguments in human evolution – is the question of how much variation a species can handle. This is particularly difficult when, in many cases, all the evidence you have on which to base your decision are bone fragments and some teeth.

Time is also an important factor: the ancient hominin species we are investigating has evolved, lived and gone extinct. For mammalian species, the average lifespan is about 1 million years. Long-enduring hominin species, such as *Homo erectus*, lived for at least 1.7 million years, and a species can change a great deal in such a long time.

Wood, in his book *Human Evolution: A Very Short Introduction*,[11] likens a hominin fossil to a still photograph in a long-distance race. Palaeoanthropologists need to decide if they are looking at several images of the same race (multiple members of the same species), or several photos from different races (multiple different species).

The argument for *Ar. kadabba*'s bipedalism hinges on a single toe bone, whose shape resembles that of later bipedal hominins. Many palaeoanthropologists are not convinced by this, especially when its large, protruding canines, the particularly ape-like sharpening of its teeth, and relatively small chewing teeth are taken into account. If more *Ar. kadabba* fossils are found in the future, we may be able to more conclusively include the species in our hominin family tree.

There is no doubt that the second *Ardipithecus* species, *Ar. ramidus* (meaning 'ground-living-root hominin'), is a hominin. For many years, it was the oldest human ancestor we knew of. In the early 1990s, a research team led by American palaeoanthropologist Tim D. White discovered numerous individuals in the Aramis region of the Middle Awash River Valley in Ethiopia.

The most impressive is Ardi, an adult female who lived there 4.4 million years ago. She would have weighed about 50 kilograms and stood at 1.2 metres tall. We know about her from some hand and finger bones, part of her thigh and lower leg, teeth and more than a hundred fragments of her skull. The skull fragments were very fragile and scattered, and it took years to stabilize and piece them together. In fact, scientists had to extract the shards of bone from the surrounding rock using a microscope and needle.

But the reconstruction effort has been well rewarded, and yielded a mosaic of ape and more recent hominin features. It showed that, while she had a small brain, Ardi's foramen magnum was closer to the middle of the skull, indicating that she may have walked upright. The top of her pelvis was broad and shallow, which would have helped to steady her gait, but the lower part of her pelvis was long like that of an ape. Around the time Ardi was alive, parts of the Middle Awash

River Valley would have been forested, with some wooded grassland. Her possible bipedalism disrupted the idea that savannas pushed hominins onto two feet.

The thickness of her tooth enamel was somewhere between chimpanzees and Australopithecines (the next genus on the human evolutionary tree), and while her canines were worn down at the tips, her chewing teeth were smaller than later hominins.

Her hands wouldn't have been able to manipulate tools in the way of later hominins and she had a widely divergent opposable big toe, which would have been good for climbing rather than walking. But an *Ar. ramidus* individual from Gona in Ethiopia, dated to between 4.6 and 4.3 million years old, had a foot that was better adapted to walking, showing that there was some variation within the species.

Some palaeoanthropologists think that *Ar. kadabba* should be combined with *Ar. ramidus* to form one large group. Other researchers go even further and say that all of these early hominins, from *Sahelanthropus* to *Ardipithecus*, should all be classed as a single genus, or even one species. It ultimately comes down to how much variation they are prepared to recognize in a single species.

How to tell what hominins ate

What a creature eats or ate tells you a lot about it. Its diet points to the environment it existed in – it can't have subsisted on fruit if it lived in a desert, for example. It suggests the cognitive complexity required to obtain the food and prepare it. Some foods have to be cooked, which speaks to the ability to forage, wield or even start fire. An animal's diet is also linked to how much of its time was occupied in the search for food. The giant panda, for example, survives on bamboo, and to get enough energy to sustain itself, it spends the overwhelming majority of its waking hours foraging and eating.

When it comes to extinct hominins, their teeth contain clues about what they ate, and from their diet we can infer a fair amount about their behaviour and lives. Many careers in palaeoscience have focused on hominin teeth.

Carbon in plants

Plants have different ratios of carbon isotopes, specifically carbon-12 and carbon-13. These are both stable isotopes of carbon, unlike carbon-14, which is radioactive and used to date organic matter. Scientists divide diets into C3 and C4 foods. C3 and C4 plants both have less carbon-13 compared to the natural atmospheric levels, but their ratios of C-12 to C-13 are different. These ratios continue up the food chain, so a predator that ate a C4-loving creature would have more C4 in its tooth enamel. A C3 diet includes fruit and leaves, food which would most likely be found in a woodland or forest. C4 plants, by contrast, include grasses, seeds and tubers, which would be more abundant in grasslands. By examining a hominid's tooth enamel, scientists can suggest what it ate and possibly where it lived.

First, there is the shape and composition. Hominins' cusped and relatively large premolars and molars allowed them to chew a variety of harder foods, expanding the possible habitats they could live in. Similarly, the thickening enamel on hominin teeth protected their gnashers from fracturing when chewing, say, nuts. Canines also get smaller and smaller as we progress along the human lineage, but scientists think that this is more likely to do with social behaviour than eating – although that does not rule out the possibility that shrinking canines could also point to a change in diet. Male apes, for example, have much bigger canines than females. Male gorillas have very large canines, but they mostly eat fruit and plants (some sub-species have a taste for ants). This has led scientists to argue that their intimidating teeth are mainly for competing with (and injuring) other males.

Second, the surface of teeth shows micro-wear, which is a bit like being able to tell whether you washed a window with a soft cloth or a pot scourer. Tubers (plants that mainly grow underground, such as potatoes) are unearthed covered in soil, and if a hominin ate a lot of them, the grit would have created a specific pattern on the surface of their teeth. Hard nuts, on the other hand, will leave impact craters.

Third, isotopic chemical analysis offers direct evidence of what our hominin ancestors ate. Isotopes are variants of the same element. For example, there are twenty-two isotopes of carbon. They all have six protons, but have a varying number of neutrons. Isotopic chemical analysis involves measuring the isotopes of various elements, including oxygen, carbon, nitrogen and zinc, and comparing the ratios to living animals whose diets we know. With isotopic chemical analysis, scientists can point to the types of food that hominins ate – such as whether they consumed more grasses than fruit – or other creatures that ate these plants.

PART 2

EARLY HOMININS

4 THE SOUTHERN APE

Millions of years ago, the landscape looked very different to how it does today. Throughout Earth's history, its polar ice caps have frozen, thawed and frozen again. They have trapped millions of cubic kilometres of fresh water in their icy depths, and when the climate heated again, this water flooded the oceans. Innumerable forests have risen up from saplings, rotted into mulch and given way to grasslands. The continents themselves have shifted, raising mountains and buckling landscapes, while volcanoes have covered the land with boiling, mineral-rich lava and ash.

The changeable landscape has very important implications for fossil-hunting. When creatures begin the long process of fossilization, they mostly become trapped in a sediment layer (although there are notable exceptions, such as creatures that fall into bogs and become naturally mummified). Layers of soil, ash and mud cover their bodies, and the chemical composition of these layers can make them easier to find and date. There are two major regions for early hominin fossil-hunting – South Africa and eastern Africa – and this is mainly due to their fortunate geology.

South Africa's 'Cradle'

In South Africa, about 50 kilometres north-west of the country's economic hub Johannesburg, there is a treasure trove of hominin fossil sites. Collectively referred to as the 'Cradle of Humankind' by the government, the area, with its green rolling hills, was listed as a UNESCO World Heritage Site in 1999. UNESCO, however, calls it 'Fossil Hominid Sites of South Africa', rather than bestowing the somewhat contested title of 'cradle' upon its undulating landscape. It only covers about 10 kilometres by 15 kilometres – a tenth of the size of London – but it is one of the richest hominin sites in the world. Palaeoanthropologists have found numerous early hominin skeletons (which are in fact remarkably rare globally), artefacts and animal remains at the sites, which have contributed greatly to our understanding of human evolution.

About 2.3 billion years ago, that area was a large, warm inland sea, home to a burgeoning community of coral reefs. The calcium and magnesium in the ocean and marine life laid the foundations for the dolomitic rock that forms part of the Cradle's geology, and is what has made it such a wellspring of discovered fossils.

There are two reasons for this. One, limestone (which is calcium carbonate) is often found with dolomite, and in the nineteenth century, lime (calcium oxide, a derivative of limestone) was a much-traded commodity in South Africa, where rapidly developing colonial communities used it as a mortar to stick bricks together in buildings. Limestone mining was big business, and in fact some key hominin fossils, such as the Taung Child (*Australopithecus africanus*) and Little Foot, were discovered because miners blasted dolomite with dynamite in search of limestone.

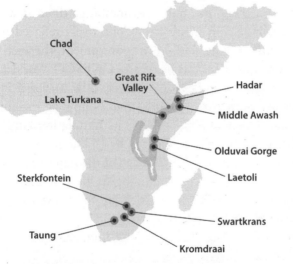

The African continent has yielded a treasure trove of hominin fossils. Thanks to eastern and South Africa's special geology and geography, among other factors, these regions are rich sites for understanding early human evolution.

The second reason for the Cradle's wealth of uncovered fossils is that water can dissolve limestone. This erosion forms cavities and caves within the rock. For many unfortunate creatures, a wrongly placed foot meant a fall into vertical shafts that water had gnawed into the rock underfoot, and a plummet into inescapable subterranean caverns below. For example, at Malapa, a rocky site populated by the odd tree and surrounded by the Cradle's grassy hills, there is a cave which contains hominin and animal fossils, none of which have the telltale signs of a carnivore's teeth. There are more than twenty-five species represented among Malapa's remarkable hoard, and these creatures most likely fell to their death through treacherous chasms millions of years ago.

But the Cradle's singular geology also makes it a very difficult site to date. These caverns and shafts can collapse in on themselves, and debris, bones and a myriad rocks from various sedimentary layers collect in pockets. They form a cobbled-together rock called 'breccia' (which comes from Latin and literally means 'rubble'). Such a mismatched collection makes dating tricky in this area. In many instances, scientists focus on the sediment closely associated with the fossil itself, so that they don't accidentally date other rocks that solidified around the specimen.

However, as techniques and technology improve, so does our ability to date these tricky sites. In a 2019 paper in *Nature*,[12] scientists described a way of constraining possible early hominin dates. During wet periods in the climate's history, rain water seeped through the bedrock, leading to the formation of flowstone layers in the underground caves and caverns in the Cradle. These layers are sheets of carbon minerals, and can be fairly accurately dated using uranium-lead dating. Yet wet conditions are not ideal for fossil formation, meaning that petrified remains in these fossil-rich locations belonged to creatures that were most likely alive during a dry period. The scientists identified times when the Cradle was likely more arid, effectively constraining when they walked the Earth.

Riches in eastern Africa

Fossils found in eastern Africa do not have the same dating problem as those discovered in South Africa, and are often used as a reference for the southern sites' fossils. Along the East African Rift, the continent of Africa is being torn apart like a piece of paper. The process first started between 22 and 25 million years ago, and some scientists think that in a few more million years eastern Africa could separate from the mainland entirely.

The rift valley runs like a 'Y' which swallowed a circle: the top branches sit in the Red Sea, cradling either side of the Middle East, before running down through Ethiopia where it splits into eastern and western branches, before reuniting in the south. In total, the East African Rift covers thousands of kilometres, and runs through several countries, including Ethiopia, Kenya and Tanzania, all of which have numerous significant hominin fossil sites.

The rift, which is the largest seismically active rift on Earth, is home to active and dormant volcanoes. While the tectonic plates were ripping the land apart, lava from active volcanoes filled valleys in the rift, covering the exposed surface. Their ash also rained down, and over time created layers of volcanic rock. In some places, the rift is so deep that molten magma from the Earth's core seeped to the surface. Layers of volcanic rock are called tuffs, and each tuff is quite distinct and can be dated radiometrically. When a fossil is sandwiched between two tuffs, it is suspended in time, with the bottom tuff layer bookending its oldest age, and the newer tuff delineating its youngest possible age.

The fecund Awash

The Awash River rises west of Ethiopia's capital Addis Ababa and our ancestors have lived in the Awash Valley for millions of years. The semi-arid area is one of the richest sites of hominin fossils in the world, and home to some iconic hominin individuals. In 1980, part of it was named a UNESCO World Heritage Site. An adult female *Ardipithecus*, called Ardi, lived there 4.4 million years ago; for many years, she was the oldest known human ancestor. The Awash Valley was also home to 3.2-million-year-old Lucy, the first known member of *Australopithecus afarensis*. Scientists have been working at the site since 1938.

The mineral content of volcanic rock can vary significantly between tuffs, with some layers weathering more speedily than others. In eastern Africa, there are areas in which weathering and the shifting Earth have exposed fossils to the surface. Some of the world's greatest hominin fossils from eastern Africa, such as Ardi, were discovered because of exposed bits of bone. In 1994, when he was still a student, renowned Ethiopian palaeoanthropologist Yohannes Haile-Selassie identified two of Ardi's hand bones on the surface of a deposit at Aramis in the Middle Awash in Ethiopia, and that led to the team of palaeoscientists finding more of her bones.

Scientists have discovered many fossils in eastern African sites, which together span millions of years. The variety of animal species creates a snapshot of which creatures cohabited with hominins, and gives clues about the environment and landscape that our ancestors lived in. At some sites, such as Olorgesailie in Kenya and Olduvai Gorge in Tanzania, palaeoanthropologists uncovered tools and artefacts that point to early hominin and human cognitive development.

Bias in palaeoanthropology

We use fossils as signposts in the story of human evolution, and in the last few decades palaeoanthropologists have become increasingly aware of the bias in the fossil record, as well as their own bias, and are now trying to correct this.

There are physical realities that distort our understanding of the human narrative. For one thing, we seldom find whole skeletons. Some parts of a hominin skeleton are simply more durable than others, so we find more enamel-encased teeth than fingers. Hands and feet are more likely to be scavenged or washed away than, say, a skull. A muscle-clad thighbone may also be more tasty to a predator than a bony skull. What this means is that certain parts of the skeleton are better represented in the fossil record, and so we place more emphasis on their evolution because we understand them better. Also, size actually does matter when it comes to fossilization, and larger creatures with bigger body parts are better able to endure the transformative process.

Species which had larger population numbers that were spread over a greater geographical area also had a higher chance of becoming fossils, whereas smaller groups were more likely to disappear without a trace.

Some environments are also more conducive to the preservation of fossils. The fortuitous geology of eastern and South Africa has preserved skeletons and artefacts for millions of years, but – as the discovery of *Sahelanthropus tchadensis* in Chad proves – there were hominins or possible last common ancestors in other places in Africa too. It is very likely that there were other species living in the region and when they died their bodies were eaten by scavengers and their bones eroded into dust.

Some places are also easier to access, whether due to the physical landscape or the political one. For example, hominin fossils have been uncovered in North Korea, but it is difficult for international palaeoanthropologists to get there. Additionally, in some parts of Asia, palaeontological finds are often described in local journals in the vernacular, putting these discoveries out of reach for non-native-speaking scientists.

Also, just because fossils are found at a certain place does not mean that that is where the hominins lived – or even where they died. A flood could have swept up hominin bodies, depositing them at a specific site. Predators could have brought bodies back to their lairs.

Another important possible bias was introduced by Charles Darwin in his 1871 book, *The Descent of Man, and Selection in Relation to Sex*. In that work, he noted that modern humans more closely resembled African apes than their Asian counterpart, the orangutan. Thus, Darwin theorized, humans most likely evolved in Africa. This supposition has guided palaeoanthropology's focus on Africa as the birthplace of humans. So far, the fossil record backs up Darwin's speculation, but it is possible that the human story is more complex than we currently think it to be.

Today, scientists recognize bias as an insidious influence that can distort their findings, and both individuals and institutions see it as something to be identified and corrected. But a century ago, that was not the case. In the early days of the discipline, scientists' own biases about race and the superiority of certain races coloured their interpretation of scientific facts. In 1925, when Australian-born South African scientist Raymond Dart proposed that he had discovered an early human ancestor in South Africa, the scientific community was sceptical of his findings, in part because many of its members did not believe that humans could have evolved in Africa as opposed to 'intellectually superior' Europe.

A forty-year hoax

Although Charles Darwin had suggested in the late nineteenth century that humans evolved in Africa, he didn't have direct evidence to support his theory. Neanderthal and early modern human fossils had been found in Europe, and Dutch palaeoanthropologist Eugène Dubois had discovered a *Homo erectus* skull-cap in Trinil, Java. In 1907, a worker at a sand mine in Mauer, Germany unearthed the first example of what would be named *Homo heidelbergensis*.

But in 1912, a British amateur archaeologist, Charles Dawson, perpetrated one of the greatest scientific hoaxes in palaeoanthropology and poisoned our understanding of human evolution for decades. He claimed to have found pieces of the skull of a human ancestor in a gravel pit in Piltdown, East Sussex. Known as Piltdown Man, the 'hominin' was considered the 'missing link' between humans and apes.

Although many scientists disputed the veracity of Dawson's claims at the time, it became the prevailing wisdom about human evolution – that modern humans had evolved in Europe. For more than forty years, the Piltdown Man stood in the way of progress and understanding our origins, before being debunked in 1953. New evidence suggests that Dawson was the sole perpetrator, and planted human skull fragments and an ape-like lower jaw, along with teeth and some tools, and purposefully aged them.[13]

What this meant practically is that actual discoveries, such as the *H. heidelbergensis* fossils in Germany and other key fossils that were discovered later, were, at least initially, dismissed.

Australopithecus africanus

In 1924, a box of fossils from a limestone quarry in Taung, South Africa was sent to Raymond Dart at the University of Witwatersrand in Johannesburg. The fossils included the face and skull of a child, which contained a mixture of milk and permanent teeth. There was also a brain cast, created when limestone sediment had filled the brain cavity. Dart specialized in brains, and noticed immediately that the brain was quite human-like, even if other parts of the skeleton weren't.

The child's brain had been relatively large, and the foramen magnum in the skull showed that it had most likely walked upright – even though it was definitely not a modern human, but a human ancestor. Dart published his findings in influential journal *Nature* in 1925, naming the new species *Australopithecus africanus* ('southern ape from Africa').[14] He was met with widespread scepticism, in part because Piltdown Man suggested that Europe was the birthplace of humanity. No other fossilized hominins have been discovered at Taung since the small hominin child, and one theory is that the then-three-year-old was snatched up by an eagle and taken to its nest about 2.3 million years ago.[15]

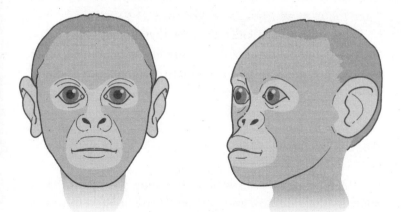

The 'Taung Child', described by Raymond Dart in 1925, was the first hominin fossil discovered in Africa.

Ultimately, Dart was vindicated when, a decade later, more Australopithecine individuals were discovered in South Africa's 'Cradle of Humankind'. The fossil record now contains many *Au. africanus* specimens, but the assemblages are often incomplete. However, when considered together, we have a good idea of *Au. africanus*' morphology.

These hominins lived between 3.3 and 2.1 million years ago in South Africa. They were not particularly large, with the males reaching just under 1.4 metres and weighing about 40 kilograms. Their brains were also small compared to a modern human's, ranging from 428 to 625 cubic centimetres, but greater in size than *Ardipithecus ramidus*, for example. *Ar. ramidus* brains were 300 to 370 cubic centimetres, which is about the same volume as a coffee mug. *Au. africanus*' brain would have been almost double that.

Au. africanus' teeth were also a lot smaller than those of great apes, but still larger than modern humans' teeth. But its hands had a blend of human and non-human ape characteristics. Its fingers were relatively curved and it had particularly strong arm muscles, indicating that it spent a fair amount of its time in trees. But its thumb and wrist suggest that it could grip objects between its fingers and thumb with precision. Its pelvis and leg bones also strongly suggest that *Au. africanus* could walk upright.

In terms of their social lives, an interesting 2011 isotope study[16] investigated the chemical differences in the teeth of *Au. africanus and P. robustus* individuals. The food and water animals eat and drink in their formative years leaves identifiable mineral traces in their teeth. The study found that *Au. africanus* females were more likely to leave their birthplace and join families further away. This social organization differs from the harem model of gorillas, for example, in which males are the ones to leave the family group.

The Taung Child is pivotal in the story of human evolution because it was the first hominin fossil discovered in

Africa. It was a vital piece of evidence that pinned the human story to Africa, and over time that narrative has shifted to the point that most of human evolution is now contained in the continent. Importantly, Dart's discovery – and later that of his colleague British-South African medical doctor and palaeontologist Robert Broom – seeded the notion that hominins once lived and thrived in Africa, and that humanity could, possibly, have originated there, paving the way for later discoveries that showed that that was, in fact, the case.

What is an Australopithecine?

Dart coined the name *'Australopithecus'*, which means 'southern ape', and in his seminal 1925 *Nature* paper described it as an 'intermediate between living anthropoids and man'. Today, it is widely considered an intermediate between earlier hominins, such as *Sahelanthropus tchadensis* and *Ardipithecus ramidus*, and later *Homo* hominins.

Australopithecus is a genus of hominin species, and there is much wrangling among palaeoanthropologists about which species should be included in the genus, whether they are in fact distinct species, and how they relate to the hominins who preceded and follow them.

This much they will agree on: species in this group lived between about 4.2 and about 2 million years ago in Africa; they were bipedal a lot of the time; they had large chewing teeth with thick enamel; and they had relatively small brains. Species within this genus are mostly defined by how they relate to the apes that came before them, and later *Homo* species, with sparse details on how they would have related to each other.

It is difficult to draw a direct, scientifically robust link between different species and genera. One of the reasons is that while we are continually adding to the hominin fossil record and deepening our understanding of hominins, the record itself is still relatively scant.

Homoplasy is another (rather large) complication in our desire to sketch a simple story arch for humanity. Homoplasy, also known as convergent evolution, occurs when two lineages or species evolve the same trait, even though they do not have a recent common ancestor. For example, bats, birds and insects all evolved wings independently. What this means in our human evolution story is that different Australopithecine species could have evolved more 'human-like' traits independently, without reference to each other.

Nevertheless, for about 2 million years Australopithecines roamed Africa. They were distinct from the hominins that came before, and had a mosaic of archaic and modern features, which were harbingers of what *Homo* hominins would display many years later.

The earlier Australopithecines: *Au. anamensis*

Since Raymond Dart described and named the Taung Child in 1925, palaeoanthropologists have discovered several *Au. africanus* individuals. They lived between 3.3 and 2.1 million years ago, but *Australopithecus anamensis* was wandering around Kenya millions of years before that.

In 1965, a research team from Harvard University discovered an arm bone at Kanapoi in the Kenyan Rift Valley, near Lake Turkana, but without any other fossils or information it sat uncategorized for almost thirty years. In

1995, British palaeoanthropologist Meave Leakey established the species, *Au. anamensis*, based on specimens found at Kanapoi and Allia Bay in Kenya. The unaccompanied arm bone was included in the new species.

The name 'anamensis' comes from the word for 'lake' ('anam') in the Turkana language. The *Au. anamensis* fossils from Kanapoi have been dated to between 4.1 and 4.2 million years old, while the specimens at Allia Bay on the other side of Lake Turkana are about 3.9 million years old. More specimens have since been found in the area.

Like other species of *Australopithecus*, *Au. anamensis* had a mix of ape-like and *Homo* characteristics. It had a projecting lower face, with a small brain case. Its wrists now suggest that it often walked on its knuckles, similar to African apes, but parts of its knee and shinbone were better suited to bipedalism. Particularly, the tops of its shinbones are thicker, suggesting that it could walk upright, a supposition bolstered by the position of its ankle. An analysis of its tooth enamel, which is fairly thick, indicates a C3-heavy diet consisting mainly of fruit.

A number of *Au. anamensis* fossils have also been found in Ethiopia, specifically in the Afar Region. They have been discovered in the Middle Awash near the discovery site of *Ardipithecus ramidus* which lived about 4.4 million years ago, and so preceded *Au. anamensis* by a few hundred thousand years. In 2019, Ethiopian palaeoanthropologist Yohannes Haile-Selassie announced the discovery of a nearly intact skull in Ethiopia, dated to about 3.8 million years old.

Many palaeoanthropologists think that *Au. anamensis* transitioned into and was ultimately replaced by *Au. afarensis*, another hominin which appears to have overlapped slightly with *Au. anamensis*, both in terms of time and, in some places, geography. Other palaeoanthropologists lump *Ar. ramidus* and *Au. anamensis* together as a single evolving species.

Australopithecus afarensis

In 1973, American palaeoanthropologist Donald Johanson discovered what would ultimately become one of our most famous early hominin ancestors. Found at a site called Hadar in the Awash Valley in the Afar Region of Ethiopia, Lucy is a member of *Au. afarensis*.

The diminutive Lucy stood at just over 1 metre tall, and was named after The Beatles' song 'Lucy in the Sky with Diamonds', which the researchers played frequently in evenings during the field expedition. In Amharic, the official language of Ethiopia, she is known as Dinkinesh, which means 'you are marvellous'. Scientists estimate that she died about 3.2 million years ago; one hypothesis is that she fell out of a tree.

There were many remarkable things about Lucy: even though palaeoanthropologists had only found about 40 per cent of her skeleton, at the time she was the most complete fossilized hominin individual to have been found. At that time, she was also the oldest hominin that scientists had uncovered. Moreover, she walked upright despite having a relatively small brain, showing that hominin bipedalism came before the significant increase in brain size seen later in *Homo* species.

In 1975, palaeoscientists excavated more than two hundred hominin fossils from Hadar, representing at least thirteen individuals, including four children. Since then, scientists have found more *Au. afarensis* fossils, making it one of the best represented species of *Australopithecus* in the fossil record. *Au.*

afarensis has been found throughout eastern Africa, including Laetoli in Tanzania, Hadar in Ethiopia and at sites in Kenya.

The Hadar hotspot

The Hadar site lies along the banks of the Awash River, on the southern edge of the Afar Triangle, which is part of the East African Rift Valley. Today, Hadar is a canvas of beige-grey sandy hills against a panoramic blue backdrop. But the fossils unearthed from the volcanic layers underfoot indicate that it was once a floodplain that continuously sprouted vegetation, and that between about 3.5 and 2.9 million years ago it was home to many species. The site isn't only known for its Australopithecine and other fossils, but also stone tools most likely used by later hominins.

Palaeoscientists established the species *Au. afarensis* in 1978 to incorporate the wide spread of individuals that had been discovered. The name 'afarensis' is a nod to the Afar people and region where Lucy was discovered.

Au. afarensis were in existence from about 3.9 to 3 million years ago, making it a particularly long-lived species. (*H. sapiens*, by comparison, have only been around for a tenth of that time period.) *Au. afarensis*, like other Australopithecines, has a mixture of ape and human features. Its members had protruding faces and relatively small brains, ranging in size from 385 to 550 cubic centimetres. This is slightly smaller than *Au. africanus*' 428 to 625 cubic centimetres, but much larger than Ardi's brain case, which had the volume of a can of Coca-Cola (300 to 350 cubic centimetres).

Some members of *Au. afarensis* had bony ridges at the top and back of their skulls, where attached muscles gave them powerful chewing capabilities. Their teeth, however, showed a marked difference from their ape ancestors. *Au. afarensis* had completely lost the canine honing complex seen in African apes. Additionally, their canines were much smaller and their chewing teeth much larger than those seen in chimpanzees. The change in tooth morphology points to a different diet, which would have included harder food. Studies of carbon isotopes in their teeth enamel also show that their diet included grassland plants, such as sedges or grasses, known as C4 plants, rather than the C3 plants that are the mainstay of an ape's diet. While Lucy and her fellow *Au. afarensis* could walk upright, as shown by their pelvises and numerous adaptations in their legs, knees and feet, they were also able to climb trees. This means that they were foraging in grasslands but also had strong arboreal abilities, which gives us an idea about the environment they lived in.

The tiniest and biggest *Au. afarensis*

More than seventy years after Raymond Dart identified the Taung Child's remains in a box of fossils, another small Australopithecine was discovered. In 2000, Ethiopian palaeoscientist Zeresenay Alemseged noticed the fossilized face of a small *Au. afarensis* girl child sticking out of an eroding hillside in Dikika, a few miles away from Hadar in Ethiopia. It took five years to painstakingly excavate her from the slab of rock.

Sexual dimorphism

In modern humans, men and women look remarkably similar. There are differences in physiology, yes, but the dissimilarities are not as stark as, say, in mandrills. In these Old World monkeys, males weigh almost 3.5 times as much as the females. The males also have brightly coloured faces – with a red line running down the middle, flanked by grooved blue skin – and the more dominant the male, the brighter his colouration. Males also have multicoloured genitals (a rainbow of purple, blue, red and pink).

Similarly, gorillas are highly sexually dimorphic, with males being much larger than females and having gigantic teeth. This adaptation is linked to social dynamics and competition among males. In humans and chimpanzees, however, there are moderate differences between the sexes. Modern humans have very little sexual dimorphism compared to gorillas. The same is – we think – true for many hominin species, although it is dangerous to extrapolate assumptions back into the fossil record.

She has been named Selam, which means 'peace', but is also sometimes called the Dikika Child or Lucy's Baby, even though, at 3.3 million years old, she is more than 120,000 years older than Lucy. She doesn't appear to have been injured or attacked by predators, and may have fallen into a river or been whisked away by flood waters.

Alemseged and his team recovered an almost complete skull, torso and several limb parts, as well as a full mouth of milk teeth with permanent teeth forming behind them. Her shoulder blades and arms, along with gripping feet and curved fingers, suggest that *Au. afarensis* was a strong tree climber.

Scientists were able to determine her age (2.4 years) by counting the microscopic growth lines in her teeth, which – due to their size – pointed to her being a female. Male *Au. afarensis* individuals are larger than females and have bigger teeth.

Small children's bones are light and fragile, and they are more likely to fall prey to predators or scavengers. This makes them particularly unusual in the hominin record. Selam gives scientists a rare insight into how *Au. afarensis* developed, both as a species and individuals.

Five years later in 2005, Ethiopian palaeoanthropologist Yohannes Haile-Selassie discovered a 3.6-million-year-old partial skeleton of an *Au. afarensis* individual in the Afar Region. He named him Kadanuumuu, which means 'big man' in the local Afar language, and that is because, relatively speaking, Kadanuumuu was a giant. At more than 150 centimetres, he was half a metre taller than Lucy and was about 400,000 years older.

Kadanuumuu offers a vital data point in the study of *Au. afarensis*. He showed the vast size differences that we can expect within the species, and the high sexual dimorphism, which is lost in individuals later in the human lineage. That said, given that scientists did not recover Kadanuumuu's teeth or skull, other researchers are hesitant to include him in *Au. afarensis*.

The Laetoli footprints

At around the same time Kadanuumuu was taking his 'giant' steps in Ethiopia, other hominins were quite literally cementing their footprints into the ground at Laetoli in Tanzania. Just over 3.6 million years ago, a nearby volcano erupted, raining fine ash down on the surrounding countryside. A spattering of rain effectively turned parts of the landscape into wet cement. Numerous species – including giraffes, baboons, rhinos and, importantly for our human story, hominins – walked through the wet ash. When the volcano erupted again, it covered the tracks, protecting them from the elements and preserving the footprints for millions of years.

In 1976, scientists stumbled upon animal prints at the site, and two years later palaeontologist Mary Leakey excavated the trail of hominin footprints. The trail comprises seventy footprints, extending 27 metres, and they are thought to have been made by at least two individuals walking together and one following behind. From their spacing and gait, we know that they were moving in an upright position in a manner similar to modern humans: the heel of their foot hit the ground first, before their toes pushed off the earth. Also, all their toes pointed forward, unlike apes. Most importantly, the find put an irrefutable date on hominin bipedalism: 3.6 million years ago, hominins were walking on two feet.

Forty years later, Tanzanian and Italian researchers

discovered more footprints about 150 metres away from the original tracks. They were made by two individuals who were going in the same direction as the first group.

Many palaeoscientists attribute the footprints to *Au. afarensis*, since only fossils from that species have been found at Laetoli. However, other hominins were around at that time, so it is possible that individuals from another species were walking through volcanic ash after the rain.

5 A HOMININ PARTY IN AFRICA

From about 3.9 million years ago, we start to see a veritable bonanza of hominin species in Africa. It could be that environmental conditions promoted a proliferation of different species. It may be that there were auspicious conditions for fossilization, such as nearby active volcanoes, and so we simply know about more hominins. It might be the case that we now have so many more palaeontologists looking for the history of life trapped in the ground under our feet. Or perhaps there has always been a large diversity of hominin species, but we simply haven't found them yet.

Whatever the reason, in the last thirty years, many new hominin fossils have been discovered and there has been a proliferation of new species. Not everyone agrees that all of these newly classified species should in fact be separate ones, but this is a common argument in palaeoanthropology.

In South Africa in 1994, palaeoanthropologist Ron Clarke was looking through a box of fossils labelled 'Cercopithecoids', and identified several bones, including foot bones, that belonged to a hominin. The specimen was named Little Foot. The fossils had been blasted out of a cave in the Sterkfontein Formation, and Clarke sent assistants off to search a section of the cave, taking with them a piece of broken-off shinbone to see if they could find where it had been broken out of the wall.

They found it in two days, and so began a twenty-three-year endeavour to liberate Little Foot from the rock and describe it. The individual is almost complete, giving rare insight into the morphology of an early Australopithecine. She was an adult female, and her leg bones were adapted for bipedalism, while her arms retained tree-climbing tendencies.

Dated to almost 3.7 million years old, Little Foot is the oldest hominin specimen from South Africa, and she is almost 400,000 years older than the famous *Au. afarensis* Lucy in eastern Africa. Clarke called her *Australopithecus prometheus*, a name associated with a skull bone discovered in Makapansgat more than sixty years before. Makapansgat is part of the Cradle of Humankind in South Africa. However, many palaeoanthropologists do not recognize *Au. prometheus*, believing that Little Foot is a type of *Au. africanus*.

In 1995, researchers announced the discovery of part of an ancient lower jaw containing a few teeth in the region of Bahr el Ghazal region near Koro Toro in Chad, in central northern Africa. (In Arabic, that means 'river of the gazelles'.) What made the find special was that it was the first time an Australopithecine fossil had been found outside of eastern and South Africa. Since then, a few more jaw fragments and teeth have been unearthed, and the fossils have been dated to about 3.6 million years ago.

Given the date, many palaeoanthropologists think that the fossils belong to *Au. afarensis*, even though they're about

2,500 kilometres away from where most of the species have been found. French palaeontologist Michel Brunet and his colleagues, who discovered the fossils, say that there are differences in the jaw and that the specimens have thinner tooth enamel, indicating that it is not a member of *Au. afarensis*. They named it *Australopithecus bahrelghazali*, after the location where it was originally uncovered.

An analysis of its teeth, however, showed that *Au. bahrelghazali* mainly ate savannah foods, different from other hominins at the time. This points not only to hominins' ability to exploit whatever food was abundant, but also gives a glimpse into what the environment in that part of Chad was like more than 3 million years ago.

Kenyanthropus platyops

In the late 1990s, a team led by British palaeoanthropologist Meave Leakey discovered a collection of fossils at Lomekwi, Kenya. One specimen in particular, a distorted skull unearthed by research assistant Justus Erus, raised more questions than it answered.

The fossils at the site have been dated to between 3.5 and 3.3 million years ago, around the same time that *Au. afarensis* was wandering around eastern Africa. But this cranium, even though it was rather mangled, looked different to other Australopithecines and shared similarities with later *Homo* species. Its face was flat – the first example of a flat-faced hominin in the record – with tall cheekbones, and it had a less protruding jaw with smaller teeth. Based on this, Leakey and colleagues argued that, despite its small brain, which was similar to other Australopithecines, it fell into a completely different genus. (*Homo*, *Ardipithecus* and *Australopithecus* are examples of hominin genera.) They named it *Kenyanthropus platyops*, meaning 'flat-faced man from Kenya'.

K. platyops was not the only important anthropological discovery at Lomekwi. The rocky expanse west of Lake Turkana in Kenya is fairly large, and in 2011 American archaeologist Sonia Harmand was looking for the site where *K. platyops* was found and stumbled on something astounding: stone tools.

The researchers reported stone flakes dated to 3.3 million years ago, more than 700,000 years older than previous tools. Prior to the discovery, the oldest stone tools, found at Gona in Ethiopia, were dated to 2.6 million years ago. In the most recent find, it looks as though individuals pounded stones against a hard surface, making a collection of flakes.

Scientists assume that the tools were used by *K. platyops* because it is the only hominin that has been found at that site. But just because the two have been unearthed together is not proof that *K. platyops* made or even used the tools.

A foot, jaws and even more species

Anthropologists and archaeologists often have to extrapolate findings from a meagre amount of evidence. Such is the case of the Burtele foot. In 2012, Yohannes Haile-Selassie and colleagues announced the discovery of eight pieces from the front of a hominin's right foot (as opposed to the full twenty-six bones). They were found in the Burtele part of the Woranso-Mille palaeontological site in Ethiopia, where sediments have been dated to about 3.4 million years ago.

Although the hominin lived at the same time as *Au.*

afarensis, this foot was different: the toes faced the same direction, but the big toe was divergent (meaning that it pointed in a different direction). This indicates that the owner of the Burtele foot would have been able to grasp branches – like *Ar. ramidus*, which was living in the same region about a million years earlier – as well as walk upright. This foot would have left footprints very different to those found in Laetoli, Tanzania, suggesting that perhaps there was more than one method of hominin bipedalism.

Ethiopia's Woranso-Mille yielded other treasures in the following years. In 2015, Haile-Selassie and other researchers announced the discovery of an incomplete upper jaw and bits of two lower ones, along with some teeth. It's unclear whether these fossils belonged to the same individual, but they have all been dated to 3.5 to 3.3 million years ago. The discoverers named the species *Australopithecus deyiremeda*, meaning 'close relative' in the Afar language.

Its cheekbones sat further forward than most *Au. afarensis* specimens, it had smaller teeth, and its cheek teeth erupted at an angle instead of coming straight up in its mouth. Haile-Selassie and colleagues argue that these distinctions are enough to consider *Au. deyiremeda* a new species, but many are dubious about the classification, saying it could belong to *Au. afarensis*.

No one is sure if the Burtele foot belongs to *Au. deyiremeda*, and it remains a bit of a mystery in an already enigmatic discipline.

The most recent Australopithecines

In 1999, Ethiopian palaeontologist Berhane Asfaw and colleagues announced a new hominin discovery in the Bouri

Formation in the Middle Awash, Ethiopia. They named it *Australopithecus garhi* ('garhi' means 'surprise' in the Afar language), and dated it to about 2.5 million years ago, making it much younger than most other Australopithecines in eastern Africa. The find was surprising in a number of ways.

It had a protruding lower face and a smallish brain (about 450 cubic centimetres), similar to *Au. afarensis*, and noticeably large chewing teeth. It also had a small bony ridge (known as the sagittal crest) running along the top of its skull. Strong muscles connect the ridge to the jaw, allowing for a powerful bite and chewing abilities. What stands out is that its leg ratio – it had a relatively long thighbone – was similar to that seen in later hominins who habitually walked upright. But its arms were ape-like, with a comparatively long forearm, and it had curved fingers, compatible with tree-climbing.

However, the big surprise was other bones found with *Au. garhi*. They bore the marks of deliberate butchering: an ancient deer's jaw with three unambiguous cut marks, a shinbone that had been smashed to liberate the bone marrow inside, and an archaic horse's thighbone with the nicks of filleting. There were no tools at the site, but at the time these scarred bones were the first evidence of hominin butchering in the human lineage. Palaeoanthropologists assume that *Au. garhi* was the one using the tools, but it could have been another one of the hominin species – including *Homo* – that were around at that time. However, a decade after Asfaw's announcement, scientists pushed the date of hominins butchering their dinner back about 800,000 years. They discovered 3.4-million-year-old animal bones in Ethiopia which bore the telltale signs of butchery using stone tools.

Almost twenty years after Asfaw announced his discovery, and thousands of kilometres away, American-born South

African palaeoanthropologist Lee Berger and his son Matthew were wandering around the Malapa site in South Africa. Nine-year-old Matthew noticed a hominin collarbone sticking out of a pile of old mining rubble. Subsequent excavations uncovered four hominin individuals, including two partial skeletons – an adult female and young male – which were found so close together that it is likely that they died at the same time. Berger named the hominins *Australopithecus sediba*, with 'sediba' meaning 'wellspring' or 'fountain' in the local Sesotho language.

Au. sediba has a unique mosaic of early hominin and later Homo features. Its brain was about the same size as other Australopithecines and it had long, tree-climbing arms. However, it had a very curved lower spine, an adaptation for walking upright. Its gait would have been quite different from other Australopithecines as its feet were hyperpronated (where the ankle rolls inwards). And while its fingers were curved, its thumb was quite long, meaning it would have been able to manipulate objects, possibly even tools, although no tools were found at the site.

The fossils have been dated to about just under 2 million years old, and its discoverers believe it may have been a descendant of *Au. africanus*. Other palaeoanthropologists think that the Malapa hominins may have been part of *Au. africanus*, although this would extend the age range of the species by about half a million years.

Giant-toothed *Paranthropus*

While it seems like there was a veritable *Australopithecus* party happening in eastern and South Africa between 4 and 2 million years ago, they were not alone.

Another genus of hominin, *Paranthropus*, was living alongside *Australopithecus* – and in fact *Homo* – species for many, many years. *Paranthropus* actually means 'alongside humans', but there is, of course, debate as to whether they deserve their own genus like *Kenyanthropus* or are just rather singular-looking Australopithecines. Some researchers refer to them as 'robust Australopithecines', since they are rather sturdy in comparison. However, the jury seems to be falling on the side of them being their own special club.

Paranthropus species have remarkably large chewing teeth, so large in fact that they are referred to as 'megadonts' ('having big teeth'). For some context, the back bottom molar on *Paranthropus boisei* is many times larger than an adult *H. sapiens*' wisdom tooth. Many of the adaptations seen in *Paranthropus* are linked to its chewing capabilities. Dental microwear suggests that they had a relatively broad diet, despite its expert chomping abilities. One hypothesis is that when resources were scarce, *Paranthropus* was able to 'fall back' on hard and chewing-intensive foods. This would have allowed them to survive lean times, which is when natural selection is strongest. It also had a pronounced sagittal crest which was an attachment point for jaw muscles and gave them a truly powerful chewing ability. By contrast, their incisors and canines were relatively small. (*Au. afarensis*, on the other hand, had smaller chewing teeth and larger incisors and canines.)

Paranthropus individuals are easily recognizable: they have distinctive high, wide cheekbones and a face that looks like a tilted plate. The first specimen appeared in the fossil record from about 2.6 million years ago, with the most recent specimen, found in South Africa, dated to between 1 million and 600,000 years ago.

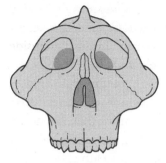
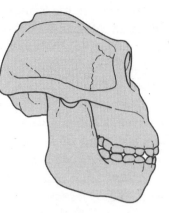

The genus *Paranthropus* lived alongside *Australopithecus* and *Homo* in Africa. They have remarkably large back teeth, with a sagittal crest running along the top of their heads.

The oldest known *Paranthropus* is *Paranthropus aethiopicus*, although it took a quite circuitous route to being included in the genus. In 1967, French palaeontologists Camille Arambourg and Yves Coppens discovered a single toothless jawbone in the Omo Valley in Ethiopia and named the species *Paranthropus aethiopicus*. Nine years later, it was reclassified as *Au. africanus*, but when researchers found a similar skull in West Turkana, Kenya, both were labelled as *P. aethiopicus*. A few more fossils have since been assigned to the species.

The original mandible was found in a 2.6-million-year-old sediment layer, anchoring the date of the species, and the youngest find is 2.3 million years old. A 2.5-million-year-old cranium in Kenya showed that *P. aethiopicus* had large chewing teeth with thick enamel, as well as an impressive sagittal crest. In the Kenyan specimen, which was discovered by British palaeoanthropologist Alan Walker, the ridge extends from the top of the head to the base of the skull. This skull is nicknamed The Black Skull because it took up magnesium during fossilization and so appears dark grey/black.

Many palaeoanthropologists believe that *P. aethiopicus* deserves its own species. Some say it should be absorbed into another group of *Paranthropus*, while others think that it should be lumped in with *Australopithecus*. So far, we haven't found any other postcranial fossils (the rest of the skeleton) belonging to *P. aethiopicus*, which could shed more light on the species and allay some of the disagreements around this portion of the human lineage.

A 'robust' hominin

While the *Paranthropus* genus includes some of the most recent early hominins that walked African landscapes, *Paranthropus robustus* was among the first hominins found on the continent.

In 1925, Raymond Dart introduced the world to *Australopithecus africanus*, and the Taung Child was met with widespread scepticism. It was only years later, after Scottish-South African palaeontologist Robert Broom uncovered numerous fossils of *Australopithecus* individuals in Sterkfontein, that Dart's assertion began to gain credence. Broom named one of the individuals *Plesianthropus transvaalensis*, nicknamed Mrs Ples, which was later classified as *Au. africanus* like the Taung Child.

In 1938, Broom made the discovery that would further support Dart's claim: a hominin that defined the genus

Paranthropus. A local schoolboy had actually found part of a jaw at the Kromdraai cave in what is now the Cradle of Humankind. The fossil fragments passed through other hands before finding their way to Broom, who hunted down its origin and found even more of the individual's remains there.

At the time, Broom noted that the fossils were more robust (heavily built) than those of other hominins, and named the species *P. robustus*. Individuals were subsequently discovered at other sites within the Cradle, such as Swartkrans and Sterkfontein, but the species has not been found outside of this rather small area. Specimens continue to be unearthed, the most recent of which was in 2020 at Drimolen Palaeocave System in the Cradle.

P. robustus lived between 2.2 and 1.4 million years ago, meaning that it would have existed alongside *Au. sediba* and early *Homo* species. Like *P. aethiopicus*, it had giant chewing teeth with thick enamel, premolars so large that they are the size of molars (premolars are the teeth behind canines), a prominent sagittal crest for powerful chewing, and a tilted-plate face. Isotopic analysis of its teeth suggest that it ate a mixture of grassland and woodland food, despite its highly specialized teeth. Its brain, while initially estimated to be about 680 cubic centimetres (which would have been the volume of two cans of beans), now seems to have been in the region of that of *Au. africanus*, at 450 cubic centimetres and 530 cubic centimetres.

For all its 'robustness', *P. robustus* was quite small, and possibly displayed a high level of sexual dimorphism. It ranged from around 24 kilograms (about the size of a Dalmatian) to 45 kilograms (a rather chubby Golden Retriever), with males reaching approximately 132 centimetres and females 110 centimetres.

Importantly for our human evolution story, bone tools have been unearthed throughout the Cradle in locations where both *Homo* and *P. robustus* have been discovered. But at one site, Cooper's D, palaeontologists have found tools and only *Paranthropus* remains, suggesting that *P. robustus* may have been the one to wield them.

The 'Nutcracker Man'

Palaeontologists have been excavating at Olduvai Gorge in Tanzania for more than a century. Its name comes from the Maasai word 'oldupai', which means 'the place of the wild sisal'. East African wild sisal (*Sansevieria ehrenbergii*) is a plant with clumps of long, spear-like leaves that grows throughout the 48-kilometre gorge.

In 1913, German palaeontologist Hans Reck discovered a hominin skeleton there – it was later dated to about 17,000 years old – and fossil-hunting duo Mary and Louis Leakey, along with other researchers, continued to excavate there for decades. They unearthed several extinct mammals and a variety of stone tools, including hand axes. These tools became known as 'Oldowan' technology, derived from 'Olduvai'. It is a specific tool-making tradition with its own unique characteristics. During that time, the Leakeys also found a hominin skull fragment and two unconnected teeth.

But it was only in 1959 that Mary really put the gorge on the human evolution map. She discovered a portion of skull sticking out of the ground, and it was labelled OH 5 (Olduvai Hominid 5). Its back teeth were so large that it was nicknamed the Nutcracker Man, because its skull and molars resembled a vintage nutcracker. Louis named the hominin *Zinjanthropus boisei* – 'Zinj' is the name medieval Muslim scholars gave eastern Africa, and 'boisei' was a nod to their sponsor, mining engineer Charles Boise. Louis argued that it didn't fit into the existing genera at the time, namely *Paranthropus* and *Australopithecus* (which he believed to be synonymous).

Louis assumed that the skull was about 500,000 years old. But in 1965, geologists used potassium-argon dating on the skull's covering tuff – the first time this technique had been used in palaeoanthropology – and discovered it was three times that age, at 1.75 million years old.

At the time, it was the oldest hominin known in eastern Africa, making it contemporaneous with *P. robustus* thousands of kilometres away. The Nutcracker Man, who was also affectionately known as 'Dear Boy' by its discoverers and their colleagues, was eventually included in *Paranthropus* and became formally known as *P. boisei*.

It had a large head with particularly giant molar teeth – even by *Paranthropus* standards – and a pronounced sagittal crest, which would have lent it powerful chewing abilities. The species' brains reached 545 cubic centimetres, which is larger than what is seen in Australopithecines. In fact, it is in the realm of *Homo*, whose brains range from about 500 cubic centimetres upwards. *H. sapiens* brains, for comparison, are two and a half times as large as those of *Paranthropus*.

P. boisei cranial remains have been found at numerous sites in Tanzania, Ethiopia, and Kenya, as well as at Malema in Malawi. However, it was only in 2013 that palaeoanthropologists discovered *P. boisei* bones belonging to the rest of the skeleton. At Olduvai Gorge, researchers found teeth and arm bones belonging to a single individual, OH 80 (Olduvai Hominin 80), which showed that the rest of *P.*

boisei was as robust as its skull. OH 80 would have weighed about 60 kilograms, leading researchers to classify him as a male and much larger than the remains of presumed female *P. boisei.*

OH 80, who died about 1.3 million years ago, is the youngest *P. boisei* individual that we have found so far. At Malema, a single *P. boisei* jawbone has been dated to 2.3 million years ago, meaning that the species endured for at least 1 million years, possibly more. *Paranthropus aethiopicus* extended back even further, with the first specimen dated to about 2.6 million years ago. During this long time span, *Paranthropus* would have lived alongside other hominins from both *Australopithecus* and *Homo*, and we do not know why both *Paranthropus* and *Australopithecus* went extinct while *Homo* thrived.

6 TOOLS FOR TINKERING

Making and using tools is not easy. Tools are implements we use to complete a task that would be difficult, or even impossible, without them. They allow us to manipulate our environment. As humans, we have an unparalleled mastery of technology, and so we often take it for granted. But to do something as simple as pick up a pencil, let alone use it, requires specific morphological adaptations. Opposable long thumbs, broad finger pads, mobile wrists and dextrous muscles all work together to create a powerful precise grip. This precision grip is the result of millions of years of evolution. However, we also need the cognitive capabilities to plan, make and use the instrument: what material to select, how to go about achieving the shape and characteristics that will allow it to achieve our purpose, and ultimately how to wield it, whether it is a pencil, an arrow or two rocks.

For many years, scientists thought tool use and manufacture were characteristically human traits. We thought that only our genus, *Homo*, had the mental abilities and morphological dexterity to manipulate our environment to achieve our desires. This was thought to be the plinth of our evolutionary success. We now know that many apes use tools – for example, chimpanzees insert long sticks into termite mounds to 'fish' for termites. A recent study even found that crows were able to build compound tools – a stick made up of three or four different pieces – to access food in a box.[17]

But a major difficulty for those researching hominin tool use came in the form of a 2016 paper in scientific journal *Nature*.[18] In it, researchers showed that wild capuchin monkeys in Brazil smashed rocks together to produce stone flakes (a process known as 'knapping'), similar to the ones found at hominin sites in Africa.

This has raised some difficult questions for the field of palaeoanthropology: if crows, monkeys and apes are able to make tools, what makes hominin tool production and use singular? A major difference appears to be what the tools are used for, the context in which they are found, and how tools evolved with our ancestors. The monkeys, for example, have not been seen using the flakes that they make, whereas in some hominin sites, animal bones have cut marks where the meat was chopped off the bone.

Nevertheless, palaeoanthropologists continue to dig away at the question of which was the first hominin to make and use a tool to manipulate their environment.

The suggestion of tool use

The oldest hint of stone tools being used is from Dikika, Ethiopia. In 2010, palaeoanthropologists described stone-tool cut marks inflicted on 3.4-million-year-old animal bones. There are parallel cut marks on the bone of a young bovid (a family which includes cattle and antelope) and others with scored Vs or shave marks on the bone, among other examples.

At that time, *Au. afarensis* was definitely living in the region, and in fact a decade earlier Ethiopian palaeoscientist Zeresenay Alemseged (who was part of the team describing the cut bones) found the Dikika Child. The finds show that 3.4 million years ago, someone used sharp rocks to scrape meat off a bone. The discoverers suggest it was *Au. afarensis*, but we have no way of knowing for sure. Also, there is no evidence that the wielder made the tools, rather than picking up a handy sharp rock. Nevertheless, the act shows planning and foresight. However, some palaeoanthropologists dispute that hominins marked the Ethiopian bones; they hypothesize that the animals could have been trampled or scratched by the sediments they were trapped in.

If someone did in fact use a stone to butcher animal carcasses at Dikika, then this find, dated to 3.4 million years, marks the start of the Stone Age. This period of hominin and human technology, in which stone was widely used as a material for implements, ran until between 6,000 and 4,000 years ago, at which point societies increasingly turned to metal to fashion their tools. The Stone Age is broken up into three periods: the Palaeolithic (Old Stone Age, which includes the first development of stone tools); the Mesolithic (Middle Stone Age, whose time period differs depending on the region but is usually meant to denote hunter-gatherer stone technology, which is characteristic of *Homo sapiens*); and the Neolithic (Late Stone Age, when people began settling into farming communities). These periods did not occur concurrently around the world, as they are distinguished by the type of technologies hominins were using at the time. To add complexity to an already convoluted field, locations often employ different terminology. In many regions, for example, scientists no longer use the term 'Stone Age', preferring specific geological periods.

The first tools

In 2011, American archaeologist Sonia Harmand stumbled upon an astonishing find. She was trawling the beige rocky hills of Lomekwi, Kenya, looking for the site where the 'flat-faced' *K. platyops* had been found, but took a wrong turn. Instead of a twenty-year-old excavation site, she found the oldest known stone tools, later named Lomekwian technology. (Stone tool technologies have their own distinctive 'flavour' and characteristics, and are named after the place where they were first discovered.)

The Lomekwi site yielded almost 150 artefacts, dated to about 3.3 million years ago. The tools include cores (the stone that flakes are removed from through 'knapping'), flakes, potential anvils and percussors (stones used as hammers). Cores record the manner and sequence in which the 'knapper' carved stone flakes off the initial rock. The flakes at the site range from 2 to 20.5 centimetres. Harmand and colleagues argue that they were intentionally made – and that there were even misstrikes recorded on the cores. The knappers' level of skill differed from that of apes, the researchers maintain, because individual objects had distinct tasks.

When the wild capuchin monkey study came out in 2016, Harmand said that the monkey-made artefacts would not look out of place on several eastern African stone-tool sites, but Lomekwian tools were different in a number of ways: they were larger than those made by the capuchins, and the rocks – consisting of basalt and phonolite – would have been much more difficult to break than the ones the monkeys used.

Who made and used the tools is another mystery. Harmand discovered the tools very close to the site where *K. platyops* was found. As far as we are aware, that was the only species in the area at the time – but that could change if new fossils are found at Lomekwi or nearby.

Early technologies

For many years, Olduvai Gorge in Tanzania yielded very few hominin fossils. It was mainly known for its remarkable collection of stone tools. Louis Leakey discovered numerous tools at the site in the 1930s, and his wife Mary created the first system of classification for the tools, although many do not agree with her categorizations.

The tools are very basic, and are often referred to as 'pebble tools'. Their creators collected river stones and then used another stone to strike flakes off them. Sometimes the flakes were used as implements, sometimes the sharp edge of the pebble became a tool. Mary divided the artefacts into groups based on their use, such as heavy and light duty, and included scrapers and choppers. This is one of the major issues that later palaeoanthropologists had with her categorization: we cannot know for certain what artefacts were used for, and whether the individuals who made them were also the ones who wielded them. Others have suggested that we refer to the tools by their characteristics, such as 'flaked pieces' and 'pounded pieces'. These names are more descriptive, but definitely less exciting.

The tools also display varying levels of sophistication. There are multiple strata at Olduvai Gorge, with the oldest layers containing the most primitive tools and more recent layers yielding more complex artefacts. In a 2021 study, researchers reported tools in strata dated to about 2 million years.[19]

Sticks, stones and bones

There is a major caveat in the study of early hominin technology: we know about stone tools because stone can survive intact for millions of years. But hominins may have employed tools made of other substances, such as wood or bone, that have not survived the ravages of time. It is possible that our ancestors had other forms of rudimentary technology that we do not know about because they have disappeared. Most chimpanzee tools, for example, are made from organic materials and will not withstand the ravages of time. The oldest known wooden tool, the Clacton Spear found in England more than 100 years ago, is about 400,000 years old.

However, there are older sites that contain Oldowan tools. One in Gona in Ethiopia has tools in strata dated to 2.6 million years old, and in a 2023 paper in the journal *Science*,[20] palaeoanthropologists describe a 2.6- to 3-million-year-old site in Nyayanga, Kenya, which contains examples of stone flaking and pounding, as well as two *Paranthropus* teeth and a butchered hippo, showing that someone was using technology to process their food.

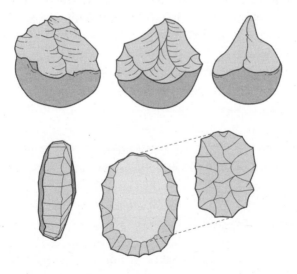

The bifacial chopper (top), in which bits of stone were removed from its edge to make it sharp, is an example of Oldowan technology. The Levallois technique (below), which appears in the archaeological record much, much later, involves chipping pieces of stone ('knapping') off a 'core' rock using another smaller rock. After creating the desired shape of the flake, a single blow separates the flake from the core rock.

Oldowan tools have since been found all over the world, including other locations in eastern Africa, southern Africa, Eurasia, and even as far as China. As *Homo* migrated out of Africa, these species took their tool-making methodology with them, and Oldowan technologies are found at many early *Homo* sites throughout Europe.

The variation in tools has led researchers to distinguish between classic and 'Developed Oldowan' technology, although Developed Oldowan overlaps with other technology systems, such as Acheulean tools which are thought to have evolved from Oldowan. In fact, all the major tool industries tend to overlap, as hominins kept old technologies that served them well while employing new innovations.

To reduce confusion, some anthropologists split stone tool industries in different 'Modes'. Pre-Mode 1 refers to the early possible tools, such as those found at Lomekwi, while Mode 1 denotes Oldowan tools and Mode 2 Acheulean. There are also more 'higher' modes that reflect the increasing sophistication of hominin and human stone technology.

Who used the tools?

Who used which tool is one of the major mysteries of palaeoanthropology. The 3.3-million-year-old Lomekwi tools were found at the site near where researchers discovered *K. platyops*, who was around at the same time. At Olduvai Gorge, there is an abundance of stone tools, and both *P. boisei* and later hominin *Homo habilis* have been found at the Tanzanian site.

Meanwhile, the 2.5-million-year-old *Au. garhi* was found along with bones that bore the unambiguous marks of deliberate butchering, but there were no tools at the site in the Bouri Formation in the Middle Awash, Ethiopia.

Without a time machine, we are unable to definitively link one species to both tool manufacture and its use. But scientists can sometimes say whether a hominin was capable of wielding tools. One of the reasons that scientists thought tool-making was the province of *Homo* was because of their relatively large brains. However, in the last few decades, the age of stone tools has been pushed back significantly and this has opened up the possibility of younger hominins, such as early Australopithecines, using tools.

To wield such artefacts, they would have had to possess the hand strength and dexterity to be able to manipulate such objects. The ability to hold a rock like a cricket ball, for example, with it sitting in the palm and fingers holding it in place, requires specific adaptations and muscle usage. You also need to be able to move it, one-handed, between the thumb and fingertips, and hold it precisely in a pinching grip. Unfortunately, hominin hands are in short supply in the fossil record, but nevertheless scientists have identified some species of *Australopithecus* and *Paranthropus* that could have possibly gripped tools.

7 WHAT IT MEANS TO BE HUMAN

We do not know how the human lineage evolved from early hominins to the more developed, more human-like *Homo*. We suspect that *Homo* emerged out of *Australopithecus*, but do not know from which specific species. In fact, we may not even have found the species that gave rise to the first *Homo*. Most of the species we've discovered were evolutionary dead ends: they lived and died, without evolving into anything else.

From about 2 million years ago, the fossil record becomes increasingly more populated – populated with specimens, variations, geographies and, according to some, many species. We have a veritable cornucopia of hominin fossils from the last 2 million years when compared with the preceding 2 million. You would think that more specimens would quiet the debates that have raged in palaeoanthropology, but that hasn't been the case. While there is more evidence, there are still gaps which researchers have filled with theories, conjecture and many differences of opinion.

The origins of dancing

Homo, unlike other mammals, habitually walked on two legs, some over longer distances than others. And one important thing that comes with long-distance walking is a repetitive, rhythmic gait. Some researchers[21] believe that this is where humans got an instinctive feel for rhythm and music, and that this was instilled in utero when pregnant mothers walked long distances. When people walk side by side, they also tend to synchronize their steps, as though walking to a beat. It is a short step from synchronized footfalls to music, the researchers argue.

In 1758, Carl Linnaeus devised the genus *Homo*, but it had just one species: us, *H. sapiens*. '*Homo*', rather unimaginatively, is Latin for 'human being' or 'man', and seemed so self-evident to Linnaeus – who didn't imagine that more species would be included in the genus – that he didn't define it further. It was only with the discovery (and acceptance) of the Neanderthals (*Homo neanderthalensis*) and *Homo erectus* in the nineteenth century that the genus got more members. In 1895, German naturalist Ernst Haeckel very ungenerously suggested that Neanderthals be called 'Homo stupidus'.

Like *Australopithecus* before it, *Homo* is not well defined. There is no firm line in the sand that delineates *Homo* from older creatures on the human lineage. Once, scientists thought that habitual bipedalism, large brains and tool-making – a mixture of behavioural and physical characteristics – defined members of the genus. But as we have seen in preceding chapters, new discoveries continue to push back the dates at which hominins walked upright and were in the vicinity of stone tools. *Homo* species tend to have longer limbs that were better suited to continuous walking, but other species, as shown by the Laetoli footprints which could have been made by *Au. afarensis*, were able to walk comfortably for some distance.

Brain size is also a complicated marker of *Homo* status. The first possible hominin, *Sahelanthropus tchadensis*, which is around 7 million years old, had brain capacity similar to that of a modern chimpanzee: about 370 cubic centimetres, or the volume of a can of Coca-Cola. *H. sapiens*' brains are almost four times that (they are on average 1,350 cubic centimetres, or just over 2 per cent of our body weight). But this evolution was not linear. About 2 million years ago, hominins' brains had grown significantly from species that emerged soon after our split from chimps, but they were still small. Later Australopithecines and *Paranthropus* had brains in the region of between 400 and 500 cubic centimetres. *P. boisei* stood out for having brains that reached 545 cubic centimetres. Early *Homo* species, such as *Homo habilis*, also had relatively small brains, ranging from under 500 cubic centimetres to about 800 cubic centimetres. But by the time *Homo erectus* appears, its brain was significantly larger, and closer to ours.

The argument about what defines *Homo* falls into a larger philosophical debate in the field of palaeoanthropology: what is a species? According to some scientists, there are many species within the *Homo* genus, each with their own distinctive morphological features. But at the other end of looked like. Creatures that appeared similar were grouped together. However, the molecular revolution meant that researchers could adjudicate who belonged in which group based on their DNA. But even with more sophisticated tools, there is still no single definition of what constitutes a species.

There are in fact at least twenty-six different published concepts of what a species is. The definition used in biology textbooks is that a species is a group of organisms in which males and females can produce fertile offspring. By this argument, two organisms from different species would not be able to interbreed, but that is often not true.

Let's say two groups of a single species become isolated from each other for a long time and evolve to the point that they are morphologically and genetically distinct; they may still be able to reproduce if they encounter each other, despite being recognized as different species.

For example, chimpanzees and bonobos shared a common ancestor between 2 and 1.5 million years ago, before the Congo River formed a physical boundary between two ape populations.[24] In 2016, an analysis of the apes' genomes indicated two episodes of interbreeding 500,000 and 200,000 years ago, showing that two very distinct types of ape were able to successfully breed.

In our own human lineage, we now know that some modern *H. sapiens* today have Neanderthal and Denisovan DNA, once again raising questions about our criteria of what defines a species.

Did a changing landscape give rise to *Homo*?

The 'savannah hypothesis' has come in and out of fashion over the years. This theory suggests that a change of habitat from woodlands or forests to grasslands effectively pushed hominins out of trees and onto two feet. A major stumbling block has been the definition of a palaeo-savannah: was it open grassland sparsely dotted with trees, or a mosaic of grasslands and trees?

Many scientists over the years have spent their careers reconstructing palaeo-environments based on carbon isotopes in fossils and soils, ancient pollen and cores from the bottoms of prehistoric lakes. When scientists found that eastern African fossils, such as *Ardipithecus ramidus*, and *Australopithecus africanus* in the south preferred woody habitats, the popularity of the 'savannah hypothesis' waned significantly – but it didn't disappear.

And then in 2017, researchers pulled together carbon isotope data from the Afar Region of Ethiopia and Turkana Basin in Kenya.[25] They analysed animal tooth enamel, fossilized soil which contained clues about what types of plants were present, and the animal fossils. They showed that almost all the animals found at the Ethiopian site of Ledi-Geraru fed on grass, indicating that there were large tracts of savannah. Ledi-Geraru is home to possibly the oldest *Homo*, dated to 2.8 million years old. About 30 kilometres away lies Hadar, where the *Australopithecus afarensis* Lucy was found. The spread of this grassy environment in Ethiopia occurred about half a million years earlier than at the Turkana Basin in Kenya. The researchers have linked the spread of grasslands to the rise of *Homo* and the extinction of Australopithecines, even though they shared a similar diet. The authors of this study acknowledge that more research is needed to prove the link, and luckily there are many palaeoclimatologists around the world trying to answer these precise questions: what was the environment like millions of years ago, and how did this impact human evolution?

The 'Handy Man'

In 1959, Kenyan excavation expert Heselon Mukiri found a single hominin tooth at Olduvai Gorge in Tanzania. His discovery was, understandably, overshadowed by Mary Leakey's *P. boisei* fossils, which were unearthed on the same expedition. But the following year, Leakey's son Jonathan found a partial skull, hand and foot bones – dated to about 1.75 million years – that obviously did not belong to giant-toothed, robust *P. boisei*, even though they would have been around at a similar time. The researchers subsequently found more specimens.

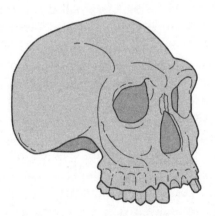

Homo habilis remains have been found at several locations in Africa, and have pinned the narrative of human evolution to the continent. However, scientists still argue about whether it should be included in the genus.

In 1964, Leakey and colleagues assigned the fossils to the genus *Homo* and gave them the species name '*habilis*', meaning 'handy' or 'mentally skilful'. This was because the remains were discovered near tools and they assumed that this species was able to use – and possibly even make – them. They argued that the species fulfilled the prevailing (in the 1960s) criteria for the definition of *Homo* – an upright posture, bipedalism and the ability to make tools – despite having a relatively small brain (600 cubic centimetres). Prior to their discovery of *H. habilis,* the oldest *Homo* species was *H. erectus.* Leakey and the others suggested that for a species to be included in Homo, it needed to have a brain larger than 600 cubic centimetres, a smooth skull which didn't have a crest like those seen in Australopiths, among other attributes.

Their announcement, like Raymond Dart's Taung Child forty years before, was met with scepticism. At that time, the human evolution timeline seemed – rather naively in retrospect – straightforward: *Australopithecus* sat at the base of our lineage; they were replaced by *Homo erectus* (specimens of which had by that point only been found in Asia), who then moved into Europe and developed into Neanderthals, who ultimately evolved into Sapiens. *H. habilis* pinned the human narrative – and the earliest humans – in Africa. As more specimens were discovered in Africa, scientists warmed to the idea of some *Homo* species arising on the continent, although it was many years before the 'Out of Africa' hypothesis – the idea that our genus originated in Africa – gained traction.

More specimens of *H. habilis* have since been discovered in Kenya, Ethiopia and South Africa, with dates ranging from 2.4 to 1.6 million years ago. But the status of *H. habilis* remains controversial. It has numerous Australopithecine features, which have led many past and contemporary scientists to recommend it be included in that genus. On the other side of this argument, there is a fair amount of variation in the different specimens that have been found, and some palaeoanthropologists suggest that *H. habilis* should actually be split into sub-species to accommodate this diversity.

We also don't know where and how *H. habilis* sits on the human lineage, whether it descended from *Au. africanus* or *Au. sediba* in South Africa; *Au. afarensis* or *Au. garhi* in eastern Africa; or if it came from a currently unknown species. Moreover, we don't know what happened next either. At one point, the Handy Man was considered an intermediate species between *Australopithecus* and *H. erectus*, but it is now more likely that it was an evolutionary dead end.

In the early 1970s, Kenya's Lake Turkana region (then known as Lake Rudolf) yielded numerous early *Homo* fossils, but we're still not sure where they fit into the human evolution narrative or even which species they should belong to.

In 1972, fossil hunters, including Kamoya Kimeu, found a hominin species by Lake Turkana in Kenya. Palaeoanthropology scion Richard Leakey described the finds, which included a nearly complete skull and some thighbones. They were ultimately dated to about 2 million years old, and originally assigned to *H. habilis*, while others believed it should be part of *Australopithecus*.

More than a decade later, the species – mostly thanks to its large brain volume of about 750 cubic centimetres – was named *Pithecanthropus rudolfensis*, after the nearby lake. Later, it was moved into *Homo* and became *H. rudolfensis*. Other specimens have since been found at other sites in Kenya, Ethiopia and Malawi, and the species' age ranges from about 2.5 to 1.5 million years ago. But even today palaeoanthropologists still argue about whether *H. rudolfensis* is a stand-alone species, or if it should be merged with *H. habilis* or *H. erectus*, or even absorbed entirely into *Australopithecus*, despite their larger brains and thickly enamelled teeth.

A taste for meat

The *Homo* genus is separated from earlier hominins by a leap in brain capacity and a reduction in tooth size. By the time **Homo erectus** begins popping up at about 1.9 million years, its brain, in some specimens, has doubled from *H. habilis* and its teeth were significantly smaller.

One hypothesis is that through using tools, *Homo* was better able to cut food – whether it be meat or root vegetables – into smaller pieces and so make them more edible. This would have widened the range of possible foods a hominin could eat. (The date of the first controlled use of fire by hominins is controversial, with optimistic estimates ranging from 1.5 million years ago, but firm evidence crops up at about 780,000 years ago.) And meat contains more nutrients and energy – required to fuel a giant brain – than plants.

In a rather unappetizing study, participants were asked to eat raw meat and vegetables.[26] Chunks of raw goat meat and vegetables were very difficult to chew into smaller pieces, but cutting or pounding them meant they could be consumed relatively easily. Since *H. habilis* has been associated with tools, some researchers infer that the species ate meat. One rather macabre suggestion was that *H. habilis* at Olduvai Gorge may have actually eaten *P. boisei*, hence the presence of both hominins and tools at the site, although there is no evidence to support this.

The Asian 'missing link'

In 1887, Dutch scientist Eugène Dubois joined the Dutch army with the singular purpose of looking for the 'missing link' between apes and man – which at that time was a rather strange preoccupation. He had been influenced by the thinking of German naturalist (and eugenicist) Ernst Haeckel (the same man who suggested Neanderthals be named 'Homo stupidus'). Haeckel disagreed with Darwin that the human lineage most likely branched off from African apes, arguing instead that the orangutan, the only non-African ape, was the best candidate for an ancient ancestor.

Dubois, with a head full of Haeckel's theories and images of orangutans, got posted to the Dutch East Indies (modern-day Indonesia) as a medical surgeon and started looking for the 'missing link'. Amazingly, he 'found' it.

In 1891, Dubois discovered hominin fossils, including a skull-cap, on the island of Java; the following year, he found a long thighbone about 15 metres away from the original specimens, which he believed to belong to the same individual. He originally called it *'Anthropopithecus'* ('man-ape'), but in 1894 dubbed it *Pithecanthropus erectus* ('ape-human that stands upright') or Java Man.

Java Man stood just over 173 centimetres tall, with thighbones that spoke of upright walking. He also had a pronounced brow ridge (thicker than that seen in modern humans), slightly projecting nasal bones and a relatively small brain capacity of about 900 cubic centimetres (though smaller than Sapiens' brains, that is much larger than those of Australopithecines).

Dubois' find was remarkably controversial. He maintained

it was the 'missing link' between apes and humans, while others thought it was an upright ape, an ancient human or an extinct side branch of the human lineage that had died off.

Its dating has also been the subject of much debate. Originally, the skull-cap was dated to about 700,000 years old, based on the group of animal fossils it was found with. In the 1980s, this was revised to between 900,000 and 1 million years old. However, in 2014, researchers dated the human-collected shells at the site to between 500,000 and 400,000 years old, suggesting that *H. erectus* may have been at the site even more recently.[27]

It has since been classified as *Homo erectus*, a pivotal species in the human story, and Java Man is the 'type specimen'. Also known as the 'holotype', the 'type specimen' is the single individual upon which scientists base a species' description. Several other specimens have been found in Java, some of which have been dated to about 1.6 million years. Scientists think that *H. erectus* lived on Java until at least 250,000 years ago.

China's hominin boom

Thirty years after Dubois' discovery, scientists began to find hominin remains in a cave system called Zhoukoudian in Beijing (then known as Peking), China: in 1921, a tooth; in 1927, another tooth; and then, in 1929, a complete skull-cap. The site has since yielded remains of about forty-five hominin individuals, some animal fossils and stone tools. The first hominin was named Peking Man, and scientists thought he was a direct human ancestor.

The lost Chinese fossils

In 1941, to safeguard them from conflict between Japan and China, scientists stored the bulk of the hominin fossils (at least forty individuals) in wooden boxes, ultimately bound for the American Museum of Natural History in New York City via ship. The fossils never made it to their destination, however, and have still not been found. Some accused the US of spiriting them away, others say the ship they were on sunk, or that the bones were even ground down for medicine. In 2012, South Africa-based palaeoanthropologist Lee Berger reported, based on a tip-off, that they may have been buried in China, on a site which is today covered with parking lots. The area has not been excavated, and the fossils' whereabouts remain unknown. Fortunately, anthropologist Franz Weidenreich had made copies of the fossils, and a lot of what we know about them is all thanks to him.

The Chinese hominins were named *Sinanthropus pekinensis* ('China man from Peking'). Like modern humans, they stood upright, with similar limb lengths. However, their brain capacities were smaller (just more than 1,000 cubic centimetres on average, compared to modern humans' 1,350 cubic centimetres). They also had startlingly prominent brow bones and wide, sharp cheekbones. There is a great deal of confusion about when they were alive, though. It would appear that *S. pekinensis* individuals inhabited the Zhoukoudian Cave several times over the course of hundreds of thousands of years, with fossil estimates ranging from 780,000 to 230,000 years ago. Excavators have found numerous stone tools at the site, but not the complex hand axes that are thought to have been used by other populations of *H. erectus*.

Between 1928 and 1937, fourteen partial hominin skulls were found at locations around China, giving rise to species such as Lantian Man (*Sinanthropus lantianensis*), Nanjing Man (*Sinanthropus nankinensis*) and Yuanmou Man (*Sinanthropus yuanmouensis*).

In 1950, the similarities between Java Man and Peking Man led scientists to group them under a single species: *Homo erectus*. As researchers described more specimens, they split them into subspecies to indicate the variation between them. So today, Java Man – the first *H. erectus* – is called *Homo erectus erectus*, while Peking Man is *Homo erectus pekinensis*. More *H. erectus* specimens were found in Asia, adding weight to the idea that humans originated in Asia.

8 A SPECIES TO TAKE OVER THE WORLD

For much of the twentieth century, Asia appeared to be the birthplace of humanity. While many hominins had been found in Africa, they bore little resemblance to modern humans. The Neanderthals in Europe and *Homo erectus* in Asia seemed much more likely candidates for our forebears, while the Australopithecines in Africa seemed closer to apes than humans. But in the mid-1970s, that narrative started to change, mainly thanks to a slew of discoveries in eastern Africa – although it took many years for the idea that humanity began in Africa to gain widespread traction.

In 1960, Louis Leakey found a fossilized skull-cap at Olduvai Gorge in Tanzania, where his team had previously found *P. boisei* and *H. habilis* specimens, albeit at a different place on the relatively large site. It had a particularly substantial brain capacity – at 1,067 cubic centimetres – and pronounced brow ridges (even for *Homo erectus*, a species known for its striking brows). Scientists dated the specimen to about 1.4 million years old. Renowned South African palaeoanthropologist Phillip Tobias suggested it belonged to *Homo erectus*.

The richness of Koobi Fora

On the eastern shore of Lake Turkana, Kenya, a ridge of sedimentary rock has yielded some of the most important hominin discoveries on the continent. The arid landscape of Koobi Fora, now part of a national park, covers 1,800 square kilometres (700 square miles), and its sediments were deposited between 4 and 1 million years ago, trapping vital evidence of human evolution in its rock. Several hominin species – *Homo habilis, Homo rudolfensis, Homo erectus, Paranthropus boisei, Paranthropus aethiopicus, Australopithecus anamensis* and *Kenyanthropus platyops* – have been excavated from its sediments. There is a large sign at the entrance, welcoming visitors to the 'Cradle of Mankind' – a similar title to that which South Africa is also vying to claim.

In 2007, researchers discovered hominin footprints preserved there in 1.5-million-year-old rock. The depressions showed that the individuals, thought to be *Homo erectus*, walked like modern people: a long stride, before landing heel first and then pushing off on their toes. More footprints were found in 2008.

But it wasn't until the mid-1970s that an African *Homo erectus* really started to become a reality. During that time, Leakey and British-born palaeoanthropologist Alan Walker found a number of hominin specimens at a site called Koobi Fora in Kenya. Notably, they discovered a piece of jaw and two partial skulls, which they assigned to *H. erectus*.

In 1975, after re-examining a fossilized jaw from Koobi Fora, scientists created a new name for the specimens: *Homo ergaster*. 'Ergaster' means 'workman' in Greek because there were stone tools in the vicinity of the fossils. Numerous specimens have since been found in Kenya, Ethiopia, Tanzania and South Africa, and they were alive between 1.9 and 1.5 million years ago.

But in 1984, Kamoya Kimeu made a truly spectacular discovery on the western shore of the lake: a nearly complete skeleton of a young *H. ergaster*. Estimated to have been between seven and eleven years old, the individual became known as Turkana Boy, or Nariokotome Boy; Nariokotome was the place where Kimeu found him. He died between 1.6 and 1.5 million years ago. He would have stood at about 1.6 metres, and would have possibly reached as tall as 1.85 metres in adulthood.

H. ergaster more closely resembles later humans than the likes of *H. habilis* and *H. rudolfensis*. Unlike earlier species, *H. ergaster*'s body had similar proportions to ours, with long legs for upright walking and shorter arms. There is quite a lot of variation in brain capacity, though: some specimens' brains were as small as 508 cubic centimetres (which would have been smaller than *P. boisei*'s 545 cubic centimetres), while others were about 900 cubic centimetres.

While *H. ergaster*'s morphology is a marked divergence from earlier hominins, its teeth are particularly striking. Its molars, premolars and jaw were much smaller than what we'd come to expect in older hominins, and this has led scientists to deduce that either its diet was significantly different from what its ancestors and their relatives had eaten, or it was now preparing food differently.

Another important suspected characteristic of *H. erectus* and *H. ergaster* was their ability to thermoregulate, although this is subject to much scientific debate. Walking or running for long distances is very energy-intensive and thirsty work. Their larger body size meant it would have taken longer for them to dehydrate, but they would have still had to develop mechanisms to cool down.

Most mammals pant to cool down. But modern humans, along with a handful of other species, have little hair on their bodies and the ability to sweat to regulate their temperature.

One theory was that, around the time of *H. erectus* and *H. ergaster* just under 2 million years ago, *Homo* species lost their fur (which would have been like wearing a fluffy blanket under the baking sun) and developed more sweat glands. There are, however, other theories for the move towards hairlessness: less hair meant fewer parasites lurking in the fur and so possibly a healthier mate; a fuzz-free face and body could have aided communication; or that perhaps we started swimming and lost the hair that would have made us cold and wet for longer.

Unsurprisingly, many scientists do not believe that *H. ergaster* should be its own species. They prefer to call it 'African *Homo erectus*'. Given how it looks, some say that *H. ergaster* is the first example of our *Homo* genus, as opposed to *H. habilis*. Some refer to the *Homo erectus* fossils found in Asia as '*Homo erectus sensu stricto*' (which means 'in the narrowest or strictest sense'), with other possible *Homo erectus* individuals from outside of Asia being called '*Homo erectus sensu lato*' ('in a loose or broad sense').

There is a fair amount of variation between *Homo erectus*, if we include specimens from all over the world. But that is to be expected in a species which endured from 1.9 million years ago to just more than 100,000 years ago, and lived in myriad environments and climates. *Homo erectus* has been found in North, eastern and South Africa, Europe, Georgia, Indonesia and China.

We also know that *Homo erectus* lived alongside several other hominins. In eastern Africa, around 1.9 million years ago, it would have coexisted with *H. rudolfensis*, *H. habilis*, and *P. boisei*, although we don't know if they ever actually interacted. Towards the end of its timeline, Sapiens, Neanderthals, Denisovans and *H. floresiensis*, among others, were also around at the same time as *H. erectus* – although possibly not necessarily in the same places.

A technology to take over the world

At about the time that *H. erectus* (or *H. ergaster*) appeared in Africa, a new form of stone technology emerges in the archaeological record. Acheulean tools, also known as 'Mode 2' technology, are significantly more advanced than the Oldowan or 'pebble' tools found at Olduvai Gorge in Tanzania and other locations in Africa. Acheulean tools, named after the site of Saint-Acheul in France where they were first found in 1859, have a distinctive, flat pear shape (called a 'biface') that can be used in implements such as hand axes. In fact, hand axes are the quintessential Acheulean technology. The oldest example has been dated to about 1.7

113

million years, and after that it is often found in conjunction with Oldowan artefacts. Hominins in Africa and Eurasia used this technology for more than a million years, during which time it saw very little variation.

To create these tools, the maker would have had to find a suitable rock or 'core', and strike flakes off it using another rock in order to fashion a flat teardrop shape that could have been used for butchering or chopping wood. How the makers actually used the tools is a matter of extensive speculation.

This technology has two important implications: firstly, it showed that there had been a stepwise evolution in cognitive ability that allowed hominins to create such technology (although not everyone agrees that stone tool mastery is a direct proxy for cognitive ability), and secondly, its utilization also broadened the user's capabilities in terms of what they could do with the implements.

The oldest known sites of Acheulean technology are Tanzania, Kenya and Ethiopia, but artefacts have been found all over the world, suggesting that the *Homo* species took the technological knowledge with them as they migrated out of Africa.

There is no definitive way to say which species of hominin first developed Acheulean technology, but we know that it does appear in the archaeological record around the same time as *H. ergaster* and *H. erectus*, and so we assume that they were the tool-makers and that these tools allowed them to migrate, adapt and proliferate in a way that their ancestors and ancient cousins couldn't.

A hand-axe workshop

In the dusty grasslands of southern Kenya at Olorgesailie, researchers have discovered hundreds of hand axes. There are so many hand axes at the site that it is sometimes called a 'hand-axe workshop', although there is no evidence it was an actual industry and processing spot. Hominins are thought to have lived there since about 1.2 million years ago. While the site is very rich in stone tools and animal fossils, the first hominin remains – parts of a skull – were only discovered there in 2003. The skull is about 900,000 to 970,000 years old, and thought to belong to *H. erectus*.

Olorgesailie is one of the best-dated sites in Africa. In 2020, scientists drilled into nearby sediments, removing a 139-metre core from the earth. This core, which was a 4-centimetre-wide cylinder, contained numerous strata, representing about 1 million years of environmental history.

In core dating, scientists extract a long cylinder of earth out of the ground. This cylinder of sediment is a bit like a multi-layered trifle, in which each layer gets older the closer you get to the bottom of the dessert. At many sites, such as Olorgesailie, we have large time lapses in the archaeological record, sometimes hundreds of thousands of years. The information in these cores can be used to reconstruct these ancient palaeoenvironments, and fill in some of the gaps.

Out of Africa: take one?

Back in the early twentieth century, it was difficult to explain how *Homo erectus* ended up in Asia. No one knew about tectonic plates, and that shifting plates could cause land masses as big as continents to move. Those scientific concepts would only be explained later in the century. The notion of a supercontinent, like Pangaea, did not exist. And yet scientists had a problem: there were lemur fossils in Madagascar and India, but not mainland Africa or the Middle East. So they hypothesized that there had been a bridging continent, Lemuria, which linked Madagascar and India, acted as a bridge for lemurs, and then sank.

In the realm of human evolution, Ernst Haeckel popularized the idea of a lost 'paradise' that was the real birthplace of humanity. In this world-view, transitional hominin fossils had sunk to the bottom of the ocean along with the land these creatures had walked.

But as scientists uncovered more fossil evidence, with firmer dates, a different narrative emerged: the *H. erectus* fossil evidence in Asia was younger than the specimens being uncovered in Africa. Today, current thought is that *H. erectus* originated in Africa, evolving from *Australopithecus* or even another early form of *Homo*. Some refer to *H. erectus* as the first cosmopolitan hominin in that it moved beyond its original continent.

In 2020, a paper in *Science*[28] suggested that presumed *H. erectus* fossils in Drimolen in South Africa could be dated to about 2 million years old, making them the oldest *H. erectus* fossils in Africa, while around the same time researchers argued[29] that a cranial fragment from East Turkana, Kenya is about 1.9 million years old. The oldest

H. erectus fossils in Java, at Sangiran, place *H. erectus* there about 1.8 million years ago.

In the early twenty-first century, a stunning fossil discovery in Dmanisi, Georgia reshuffled the human narrative again. The area around the town had been settled since the Bronze Age, making it a magnet for archaeologists. But in 1991, researchers found a mandible in an ancient sediment layer that was later dated to 1.8 million years old. By 2005, scientists had unearthed five well-preserved skulls, numerous bones and some stone tools.

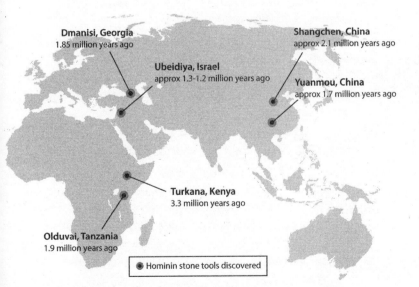

Dmanisi, Georgia
1.85 million years ago

Shangchen, China
approx 2.1 million years ago

Ubeidiya, Israel
approx 1.3-1.2 million years ago

Yuanmou, China
approx 1.7 million years ago

Turkana, Kenya
3.3 million years ago

Olduvai, Tanzania
1.9 million years ago

● Hominin stone tools discovered

It is difficult to tie tools to a specific hominin species because we do not know for certain who used them. In some cases, such as in Shangchen in China, the story is even more complicated – scientists have found stone tools, but no hominin remains.

Out of Africa 1

The theory that *H. erectus* began in Africa and migrated into other parts of the world, known as 'Out of Africa 1', is the current prevailing theory of how the species spread throughout the world. Some people believe that *H. erectus* (or the African version *H. ergaster*) was not the first hominin to make the great trek out of Africa, and that Australopithecines could have traversed the paths across the Levantine corridor and Horn of Africa, when the conditions allowed for it.

Also, neither idea precludes hominins from returning to Africa from other places, and complicating the narrative of human evolution. While such an argument may seem like sophistry, it feeds into a larger debate about the origins of modern humans. There is compelling evidence that *Homo sapiens* evolved in Africa and then spread to the rest of the world, but we now know that they interbred with other humans on their journey. Some of these Sapiens had children with Neanderthals and Denisovans (who also interbred). It is possible that earlier *Homo* migration waves resulted in populations which interbred and whose genetics are part of modern human ancestry.

The fossils' dates are controversial, but when they were discovered they were the oldest hominin remains found outside of Africa. The Dmanisi hominins showed that *Homo* had moved out of Africa sooner than originally thought. Once again, scientists dispute their taxonomy. Some say that the Dmanisi hominins are a new species, *Homo georgicus*, others say they should be called *Homo erectus georgicus*. Yet others say they should be *Homo erectus ergaster georgicus*. But for most they are referred to as the Dmanisi hominins.

Reaching up to 166 centimetres, the Dmanisi individuals would have been shorter than other *H. erectus*, such as Nariokotome Boy in Turkana, although they would have had the same long striding legs. Some of their other features seem more archaic, though: longer, more ape-like arms, and small brains (between 545 and 730 cubic centimetres).

The age range of individuals has also given scientists something to think about. They include what was most likely an elderly male, two other males, a young female and a juvenile. Interestingly, they exhibit more sexual dimorphism than seen in other *H. erectus*. But the truly remarkable part of the find was that the older man had lost all but one of his teeth well before he died. Back then, the world was a challenging place even for healthy hominins, and this man would have struggled to chew and physically find food, yet he lived for many years. Some scientists say that this points to social care, and that others would have had to help him survive. At the very least, they would have had to give him preferential treatment.

New discoveries continue to shake up the narrative of humanity. In 2022, archaeologists discovered a single hominin vertebra, dated to 1.5 million years ago, at Ubeidiya in Israel's Jordan Valley.[30] They believe it belonged to a juvenile *H. erectus* individual, who was about 155 centimetres tall when they died. This is a great deal to extrapolate from a single vertebra, but it does reinforce the idea that there were hominins outside of Africa at that time.

However, nothing in palaeoanthropology is set in stone: a 2018 paper has upset some of our certainties about the human lineage's movement out of Africa: in the paper, scientists dated stone tools in China to about 2.1 million years, using switches in the Earth's magnetic field to determine their age.[31] This is long before current wisdom thought *H. erectus* left the grassy woodlands of Africa. No hominin remains have been found with the tools, though, so we are not sure who was using them.

Each new fossil discovery shifts back the possible dates when *H. erectus* left Africa. What now seems most likely was that there were successive waves of migration, with individuals, parties and communities crossing over the Levantine corridor and Horn of Africa into Eurasia.

Why did they leave?

Homo erectus is one of the most successful species we know of. During its long existence, it colonized large swathes of the world. The current consensus is that the species began in Africa and spread to other continents. But the major question – if we assume the Out of Africa 1 hypothesis to be correct – is why.

There are a couple of suggestions, although these are only theories as we cannot truly know their motivations. An important factor is that *Homo erectus* had the ability to do it. It had long legs and could probably thermoregulate, which meant that it could journey well outside of its territory. *H. erectus* was well adapted to live in a variety of landscapes, from savannahs to woodlands. It could use tools and eat a wider diversity of foods, so the species was not limited to one location as many of its ancestors might have been. So one reason for the species' travels may simply be that it was curious and it could.

Another theory that is gaining traction is climate change. Palaeoanthropologist Rick Potts from the Smithsonian Museum in the United States and his colleagues have tracked climate variability – fluctuations between wet and dry, warm and cool – and suggest that climate variability spurred hominin adaptations. Resilient individuals who could make tools and adapt were more likely to survive, and Potts maintains that key developments in human evolution – such as bipedalism, technology and migration – coincided with climate variability. Not everyone agrees, with some researchers pointing out that hominins, who had relatively short lifespans compared to Sapiens today, would not have felt climate variability as something sudden that drove their decision-making.

A linked hypothesis is that climate variability possibly pushed other animals to migrate to more favourable environments, and that *Homo erectus* followed them. Part of human species' strength has been their resilience to adapt to changing environments, which many other animals – including their prey – did not share. A migration trigger assumes that there was one event or even a handful that pushed *Homo ergaster* out of Africa and into Eurasia. In the future, it is possible we may find that the real story is much more complex.

Similarly, we do not know what happened to *Homo erectus* more than a million years later. The remarkable traveller colonized half of the world, but suddenly disappeared from the fossil record.

In the Levant region, at least, scientists have a hypothesis: *Homo erectus* was displaced by another *Homo* species – and we know this because elephants (*Alphas antiquus*) disappeared at the same time. *H. erectus* relied on the elephants' fat, the argument goes, and when the new Homo species moved into the region about 400,000 years ago, it exploited the elephant population to extinction, effectively starving *H. erectus* out of the area or even existence.

The most recent *H. erectus* that we know of (at the time of writing) was at Ngandong on the island of Java, between 117,000 and 108,000 years ago. There were many other *Homo* species around at that time, but now there is only one: us. This is one of the great mysteries of human evolution.

Food for thought

Between 2 and 1 million years ago, we see a significant jump in hominins' brain size. *Homo habilis*, the Handy Man, was walking the African landscape 2 million years ago, with a brain just larger than a pint of beer. Fast forward a million years, and *Homo heidelbergensis* in Kwandwe, Zambia, and also around Europe, was balancing a brain double that size (about 1,280 cubic centimetres) on its shoulders. What happened to cause such a giant jump in brain size?

As is often the case in palaeoanthropology, we're not sure, but we have a number of theories. During this time, *Homo* teeth changed, becoming much smaller (especially compared to megadonts like *Paranthropus boisei*). *Homo ergaster* (the African *Homo erectus*) had teeth that were noticeably different from its predecessors, indicating that the species had undergone a significant dietary transformation.

Big brains burn a lot of energy. In modern Sapiens, our brains are about 2 per cent of our total body weight, and so need a lot of food to fuel. For some context, Sapiens have more than double the brain neurons of other primates (on average 86 billion). Our brains, containing all those neurons, consume 20 per cent of our body's energy when at rest compared to about 9 per cent in other primates.

Other large-brained *Homo* also would have needed a lot of nutrient- and energy-dense food. Some scholars argue that the use of fire accelerated *Homo*'s brain development as this would have allowed them to eat a wider variety of foodstuffs, and that there is no way they could have found the energy to fuel their big brains on uncooked food. The issue of when humanity first lit its own match is highly contested, with some saying it was from 1.7 million years ago around this brain size leap, but the earliest evidence of this is from about 780,000 years ago.

The first hint of butchery is from about 3.4 million years ago in what is now Ethiopia, showing that hominins (in that case, probably *Au. afarensis*) were eating meat, even though they were unlikely to have been able to cook it. Early hominins would have needed to be able to cut meat into smaller pieces to be able to chew it, and the earliest technology, Oldowan technology, only appears in the archaeological record 3.3 million years ago.

Earlier hominins were also likely to have smashed bones apart to access the fatty nutrients in the marrow and brains, something which some scholars say would have put them on the path of brain growth.

But starches and carbohydrates would have been key to unlocking the energy required for massive brains, and to sustain lengthy marches on long legs. Starches contain glucose, but these are often trapped in an indigestible form and need to be cooked – leading back to the unanswered question of when humans were able to roast food over an open fire.

A particularly compelling study of tooth bacteria suggests that Sapiens and Neanderthals – the largest-brained *Homo* species we know of – both had bacteria that processed sugars from starches. Since they have these bacteria, and other modern primates such as chimpanzees, gorillas and howler monkeys don't, Sapiens and Neanderthals probably inherited the bacteria from a common ancestor that had started eating starches. Scientists think that the two species split between 800,000 and 600,000 years ago, so that suggests that at least one *Homo* species was eating starches then.

An explosion of *Homo* species

For many years, there was an abundance of *Homo erectus* fossils in Asia and Africa, but a dearth of old *Homo* species in Europe. Palaeoanthropologists knew about Neanderthals and early Sapiens, but there was nothing before them.

In 1994, in the Sierra de Atapuerca hills in northern Spain, palaeontologists discovered some hominin jawbones and teeth in the Gran Dolina cavern. Since the discovery of stone tools there in the 1970s, the site had been of interest to archaeologists. They dated the sedimentary layer containing the hominin fossils to at least 780,000 years old, using biochronology and palaeomagnetism, and the species was named *Homo antecessor*. Since then, several more individuals have been discovered. However, the bones mainly belong to younger specimens, which complicates their description and comparison to other species. The name 'antecessor' comes from the Latin for 'pioneer'.

The Atapuerca hills in Spain

The early twentieth century saw the construction of a railway through the green-shrubbed hills of the Atapuerca Mountains (Sierra de Atapuerca) in Spain. Underneath the surface, encased in its ochre-brown rock, some of Europe's most important fossils lay for hundreds of thousands of years. The hills contain numerous excavation sites, including the Gran Dolina cavern and the Sima de los Huesos cave.

Archaeologists have excavated the hills since the 1960s, uncovering numerous caverns filled with hominin remains, tools, and numerous plant and animal fossils. The site has been dated to between 1.2 million and 500,000 years old. It has been listed as a UNESCO World Heritage Site, as it contains the testimony of some of the earliest people in Europe.

For many years, the site was thought to house the oldest hominin in Europe. But in 2013, Spanish palaeoanthropologists described the tooth of an infant hominin, dated to between 1 and 1.7 million years old, found at Barranco León in Spain. It usurped the title of Europe's oldest hominin fossil by up to 1 million years.

H. antecessor looks surprisingly modern, from what we can tell. There is no complete skull, only parts of the cranium, face, jaw and teeth – and what we know about the face is mainly from the remains of what scientists suspect is a ten-year-old child. Unlike *H. erectus*, *H. antecessor* didn't have striking brow bones, and it had a prominent nose and high cheeks.

Ancient footprints in England

In 2013, particularly stormy weather washed away a layer of sand on the foreshore at the English village of Happisburgh in Norfolk. Underneath was ancient sediment from a long-vanished river, and cemented into its surface: hominin footprints. Dated to more than 800,000 years ago, they are the oldest known footprints outside of Africa. Some scientists say they are closer to 600,000 years old. Once exposed, tidal erosion destroyed the footprints within two weeks, but luckily scientists took casts and 3D images of the traces as soon as they were discovered.

H. antecessor was the only other hominin species in Europe at that time (that we know of), and so is often linked to the footprints. No hominin specimens have been found at Happisburgh.

The juvenile and children's remains show the distinctive marks of butchering, leading to the assumption of cannibalism – although we don't know how this came about. Some scholars suggest that it could be evidence of inter-tribal rivalry; others propose that the younger individuals could have already been ill or dead from natural causes and were eaten to avoid the waste of meat.

Nevertheless, *H. antecessor* remains a significant species on the human lineage, even though researchers aren't sure where exactly it belongs on the evolutionary branch. Some argue that, given its relatively modern features, it could be the last common ancestor between Sapiens and Neanderthals. Others believe that *H. antecessor* branched off just before that split, and so is not directly related to us.

A possible common *Homo* ancestor: *Homo heidelbergensis*

In 1907, more than a century before a storm revealed the ancient English footprints, miners in a sand quarry near Heidelberg in Germany discovered a single chinless jawbone. German anthropologist Otto Schoetensack defined it as a new species: *Homo heidelbergensis*. Dated to about 640,000 years old, the mandible – which included most of its teeth – is the type specimen, so even though there are older and better represented examples, the species is called *H. heidelbergensis*.

Matters became complicated, as they are wont to do in the story of human evolution, with the discovery of a skull in southern Africa a few years later. In 1921, miners in what is now Zambia unearthed a hominin skull – this was even before Raymond Dart announced the discovery of the Taung Child in South Africa. The skull was found at a place called Kabwe, or 'Broken Hill', and was soon shipped to London and described as *Homo rhodesiensis*, or Rhodesia Man. (Zambia was, at that time, known as Northern Rhodesia.) Since the 1970s, the government of Zambia has been calling for the skull to be repatriated. It is currently kept at the Natural History Museum in London.

For a long time, the skull was thought to be fairly young – about 30,000 to 40,000 years old. Dating in 1974 put it at 110,000 years old. But in 2020, scientists estimated the skull to be between 324,000 to 274,000 years old, meaning that it was living at the same time as many other *Homo* species.[32]

The individual had thick brow ridges, a relatively giant brain (about 1,230 cubic centimetres) and numerous cavities in its teeth. It is one of the earliest known examples of a specimen containing dental cavities, and some have hypothesized that dental disease could have actually been the cause of death. *H. rhodesiensis* is now considered a subspecies of *H. heidelbergensis*.

In 1976, members of an expedition led by American palaeontologist Jon Kalb discovered a 600,000 year old skull at Bodo D'ar in the Awash River valley of Ethiopia. Known as the Bodo Cranium, some argue that it is a standalone species, *Homo bodoensis*. The cranium bears similarities to the Kabwe skull as well as others found at Lake Ndutu in northern Tanzania and Saldanha in South Africa. Many scholars use *H. bodoensis* as a synonym for *Homo rhodesiensis*, whose name is derived from imperialist Cecil John Rhodes, who is now strongly associated with the negative consequences of colonialism.

Specimens from numerous countries in Europe and Africa have been assigned to *H. heidelbergensis*, and some palaeoanthropologists argue that the species may also have existed in India and China. But it remains a contentious species. In 1950, *H. heidelbergensis* was absorbed into *H. erectus* (becoming *H. erectus heidelbergensis*), but these days it is usually considered its own species.

The arguments around *H. heidelbergensis* tend to be intense because it is thought to be a chronospecies, a single lineage that links *Homo ergaster* (or African *H. erectus*) with Neanderthals. Some say it could have even been a common ancestor to both Neanderthals and us. Others disagree, arguing that *H. antecessor* is the best candidate, while another group says that we haven't yet found a fossil species with the morphology expected of a common ancestor.

Part of the difficulty is that the type specimen found in 1907 is a mandible, which has limited diagnostic features. This has led some to say that European *H. heidelbergensis* individuals should be relabelled as Neanderthals, while those outside of Europe should be called *H. bodoensis* (rather than *H. rhodesiensis,*).

As *H. heidelbergensis* colonized more territory, it would have had to adapt to different terrains and climates. In Europe, for example, it would have been necessary to deal with a stark seasonality that is not present in Africa, as well as learning to cope with colder weather. There would also have been different types of animals to hunt. In order to overcome these new challenges, *H. heidelbergensis* would have had to develop new technologies and perhaps even more sophisticated forms of social interaction.

About 500,000 years ago, at Kathu Pan, South Africa, a hominin – possibly *H. heidelbergensis* or *H. rhodesiensis* – knapped spear tips out of stone. The tools are damaged along the edge, but not in a way that would have occurred if they had been used to scrape or cut. Rather, these stones appear to have been utilized as spear tips, showing that someone had harnessed spear technology half a million years ago. It is many more years before other spears are found in the archaeological record.

Far away on the other side of the equator at a site in Schöningen, Germany, researchers excavated a set of ten wooden spears, alongside numerous animal bones, and stone and bone tools (although no handles) between 1994 and 1999. Often referred to as the 'Schöningen Spears', the artefacts are between 337,000 and 300,000 years old. The animals that were unearthed along with the spears were large and showed signs of butchering. No hominins have been found with the artefacts, but *H. heidelbergensis* seems like the most likely candidate or even early Neanderthals. In 2023, scientists announced the discovery of 300,000-year-old footprints at the Schöningen site, which are the oldest footprints in the country and were possibly made by *H. heidelbergensis*.

What makes the spear find significant is that it shows the spear users were hunting, not simply scavenging their food. Also, this implies that they were able to communicate with each other and had the cognitive skills necessary to plan and hunt together. Another school of thought is that they were throwing spears, marking a shift in hominin technology.

At this point in history, there are other noteworthy technological finds that shed light on hominin cognition. In 1993, archaeologists from University College London unearthed part of a hominin shinbone at a site at Boxgrove in West Sussex. In 1995 and 1996, they discovered some teeth (lower incisors) about a metre away. The fossils are about 480,000 years old and are thought to belong to *H. heidelbergensis*. Given the length of the tibia, the individual – probably a man – would have stood at about

180 centimetres. These remains are the oldest known evidence of *Homo* in Britain. But what makes it a truly remarkable find are the other artefacts discovered at the same location.

One part of the area is known as the 'Horse Butchery Site', where archaeologists discovered the remains of a single large horse and more than 1,750 pieces of snapped flint. Researchers have been able to reconstruct numerous human activities at the site – from separating waste material to transforming parts of the horse into new tools. They also suspect that a large number of individuals were involved in the butchering and associated activities. Boxgrove shows that early humans, who may have been *H. heidelbergensis*, were able to communicate, engage in routine tasks together, and had a social life and culture.

That is not the only surprise from this era in human evolution. In 2004, at Gesher Benot Ya'aqov in Israel, researchers found evidence that hominins were controlling fire about 780,000 years ago, during the time that *H. heidelbergensis* was around. A 2022 study goes one step further and suggests that these ancient humans were eating cooked fish and discarding the fish teeth.[33]

As things stand, many palaeoanthropologists believe that *H. heidelbergensis* is an evolutionary link between African *H. erectus* (*H. ergaster*) and Sapiens and Neanderthals, and is a chronospecies. Some even think that *H. heidelbergensis* is the last common ancestor between Sapiens and Neanderthals. What we do know is that *H. heidelbergensis* was widespread, with fossils found in numerous countries, and that at the time it lived, the archaeological record starts to reflect increasingly complex and technologically capable behaviour by someone – possibly *H. heidelbergensis*.

When did *Homo* start using fire – and who ignited the first flames?

We do not know for certain when humans began controlling fires – and which species was the first to do it. Part of the problem is the definition of 'control'. It is possible that older species recognized the opportunity of fire and were able to use it if, for example, lightning set a tree alight. But that is a very different proposition to foraging tinder and starting and maintaining a fire for a specific purpose – and doing it on a continual, routine basis.

Yet the issue of fire is at the heart of human evolution. Humans are the only creatures that have learnt to control fire and to conjure it into being at will. With fire, we were able to roast foods that we otherwise may not have been able to eat, giving us access to a nutrient- and calorie-rich diet. Fire would have provided protection from predators, and its warmth would have allowed us to live in cold places.

Humanity's use of fire likely began with 'fire foraging', which only requires the ability to recognize fire. Wildfires are not only useful for their flames: in their wake, hominins could, for example, scavenge burnt remains or identify birds' eggs. The next challenge would have been to extend fire both in space (moving it from different places) and time (making it last longer). In cold places, fires would have had to be maintained through winter, and in warmer places, they would have had to be fed in the wet seasons.

But it is difficult to find these ephemeral traces in the sediments, when hearths can sometimes be as thin as a few millimetres. It is even more difficult to conclusively link the presence of burnt material to wilful fire use.

There is some evidence for early fire use in East Turkana

and Chesowanja in Kenya, from about 1.5 million years ago,[34] and there are suggestions of ancient fire at the caves of Swartkrans and Wonderwerk in South Africa between 1 million and 500,000 years ago. But at Gesher Benot Ya'aqov in Israel, archaeologists have uncovered layers of burnt materials, dated to about 780,000 years old, suggesting that whoever lived there was habitually igniting, tending or bringing fire to the same place.

Meanwhile, at Zhoukoudian in China (where Peking Man was found), the sediments, which are thought to be between 500,000 and 200,000 years old, contain burnt and unburnt bones, even though there is no direct evidence of fire.

From about 400,000 years ago, signs of fire become much more common at sites in Europe, the Middle East, Africa and Asia. Around this time, *Homo* species developed Levallois technique and there have been instances of hafting (or gluing) tools and artefacts together using resin, which would have required fire.

Early humans' fire-making capabilities usually revolve around whether they were able to start fires themselves, however some scholars flag that fire maintenance and foraging from neighbours' hearths may have been more important. Some researchers highlight that while kindling fire – rubbing two sticks together – is considered cognitively advanced, it is arguably less difficult than making Mousterian tools or hafting objects together with resin. Striking two appropriate rocks together is more advanced (the 'strike-a-light model' that uses flint and pyrite to start a fire), and we start to see that much later in the archaeological record, possibly from about 85,000 years ago.

A 2023 study suggests that Neanderthals invented the strike-a-light fire-making technique, while different Sapiens in Africa may have independently devised fire drills, where a stick is drilled into a piece of wood to start a fire and which is still the most-used hunter-gatherer fire-making technique.

At sites in South Africa, early Sapiens warmed stones to improve their flaking properties from about 164,000 years ago.[35] Such heat treatment would have expanded the possible artefacts they could engineer. New evidence from a site in Valdocarros in central Spain suggests that someone was able to control fire by at least 250,000 years ago. And from about 100,000 years ago, fire becomes ubiquitous in the human record.

Homo mysteries

While *Homo* is the best represented genus in our fossil record, there are still many gaps that we have filled with suppositions and theories. New fossil discoveries – and reanalysis of old fossils using new technology – continue to disrupt our human evolution narrative.

Before the beginning of the twenty-first century, scientists thought that only Sapiens had managed to travel beyond South East Asia. To go further required the use of boats or rafts, a technology considered the purview of our species. But a remarkable discovery in a cave on the island of Flores in Indonesia in 2003 changed that. In the Liang Bua, which means 'cool cave', researchers found the partial skeleton of a tiny human. Buried under a thick layer of ash, they initially thought it was a child. However, the individual's wisdom teeth had erupted, indicating that it was an adult, even though it stood just over a metre tall. The fairly complete skeleton (for palaeoanthropology) is thought to have belonged to a thirty-

H. floresiensis, for its age, has a peculiar mixture of archaic and modern features: it had a small brain (about 410 cubic centimetres, which is a lesser size than many Australopithecines), but its long, low cranium is more akin to *H. erectus* or even *H. sapiens*. While its toes all faced forward, its feet were noticeably long for *Homo*, and so it would have had a different gait to us. Initially, researchers suggested that, given the small brain size and stature, *H. floresiensis* may have been Sapiens with some form of birth defect or congenital disease. However, there are no known diseases that would cause these sorts of characteristics, and other discoveries on the island suggest that older hominins may have had some of their traits. But much of the debate is extrapolated from a single skull, making it difficult to say whether the skull is indicative of the species or a single particularly unwell individual. Also, importantly, *H. floresiensis* had no chin; chins are a singular feature only seen in Sapiens.

This species' evolutionary trajectory is a mystery. Some think that it evolved from Indonesian *H. erectus*, but other scholars point to their similarities with *H. habilis* and even *Au. sediba* in eastern and South Africa. How these small people ended up in Flores is one of palaeoanthropology's mysteries, but it seems likely that the species evolved on the island rather than elsewhere, as its location may explain both its relatively rapid evolution and short stature. The 'island rule', proposed in 1964, suggests that large-bodied species tend to evolve to shrink when they settle on an island. The opposite also happens: when a small-bodied animal finds itself on an island, it's inclined to evolve to be larger, such as, for example, the dodo. In 2019, some researchers modelled *H. floresiensis'* evolution and found that the hobbit very likely rose out of the 'island rule', shrinking from the size of *H. erectus* in about 360 generations.

The mystery of Maba Man

In 1958, farmers found an ancient cranium near a Chinese village called Maba in Guangdong Province. The remains included a skull-cap, parts of the upper face, and sections indicating beginnings of the nose. Labelled Maba Man, scientists suspect he was a transitional hominin, somewhere between *H. erectus* and Sapiens. Some think that the individual may have been an Asian extension of *H. heidelbergensis*. If reconstructed correctly, Maba Man would have had a large brain of about 1,300 cubic centimetres, which puts it on a par with Neanderthals and Sapiens. We're not sure of its age, but it ranges from 300,000 to 130,000 years (which is when Neanderthals were in Europe). What makes Maba Man particularly interesting is that he had derived traits of both Neanderthals (a prominent nose and thick skull) and Sapiens (the forehead). Some scholars believe that this is due to convergent evolution, in which traits in different species evolve separately, rather than Neanderthals actually making it that far east.

A few years after the discovery of *H. floresiensis*, researchers uncovered another diminutive human thousands of kilometres away on the island of Luzon in the Philippines. When the first remains were found in 2007, they were thought to belong to a modern human. But after the unearthing of more specimens in 2019, they were placed into a new species: *Homo luzonensis*. Scientists think that the remains belong to at least three individuals, and they've been dated to about 50,000 years old.

Like *H. floresiensis*, *H. luzonensis* displays a mosaic of Australopithecine and *Homo* features: small teeth, but curved finger bones and remarkably Australopithecine feet. However, the fossil record for the species is so sparse that it is difficult to say how the species walked, how tall it was and whether it exhibited the same dwarfism seen in *H. floresiensis*. Like the hominins of Flores, scholars suspect that *H. luzonensis* may have descended from Asian *H. erectus*, but that remains speculation.

Rising Star hominins

In 2013, cave divers in South Africa's Cradle of Humankind stumbled upon what would end up being a hominin treasure trove. The Dinaledi Chamber in the Rising Star Cave can only be accessed via narrow chutes, sometimes as little as 25 centimetres high (less than the length of a standard ruler). At the time of the discovery, expedition leader Lee Berger put out a call for 'skinny anthropologists, biologists, cavers, not afraid of confined spaces', ultimately choosing a team of women to explore the cave system. They were dubbed 'underground astronauts'.

The expedition yielded more than 1,550 specimens belonging to at least fifteen different individuals. More remains have since been found in a separate part of the cave system. The new species, named *Homo naledi* (which means 'star' in the Sotho-Tswana local languages), was initially thought to be between 2 and 1 million years old. This was an estimate based on their morphology, as the South African limestone cave systems are notoriously difficult to date, and they exhibited numerous archaic features that were more at home on Australopithecines, but also a range of *Homo* traits.

H. naledi had a small brain (approximately the size of a fist), curved fingers, and ancient-looking shoulders and hip joints. But its wrists, hands, legs and feet resembled later *Homo* species. Nevertheless, *H. naledi* was thought to sit near the base of the *Homo* genus, an early ancestor on the human lineage. But in 2017, scientists announced that the hominin had actually been alive a few hundred thousand years ago, between 335,000 and 236,000 years ago. This would have made *H. naledi*, which has only been found in the Cradle, a contemporary of early Sapiens and also the individuals living near Kabwe in Zambia (sometimes known as *H. rhodesiensis* or *H. heidelbergensis* or *H. bodoensis*).

Berger and colleagues initially suggested that *H. naledi* may have intentionally disposed of its dead by depositing them in the Dinaledi and Lesedi Chambers, but other scholars question whether a creature with such a small brain relative to its size could have had the cognitive capabilities for such ritualistic behaviour. In 2023, Berger and fellow researchers published a series of papers suggesting that *H. naledi* not only buried its dead, but also had mastery of fire and etched on the walls of the burial chamber. So far, only one of those studies – about intentional burial – has been peer-reviewed, and it was universally criticized by the reviewers, who skewered many of its assertions.

9 OUR CLOSEST COUSINS

When the first Neanderthal fossils were found in 1829, we had no idea that ancient species of humans existed. People assumed that humans had appeared fully formed and looked the same as a biblical Adam and Eve. Swedish taxonomist Carl Linnaeus had described us as *Homo sapiens*, but did not give a specific description of the genus's traits. He listed 'varieties' of humans, based on their geographic locations, skin and hair colour, and assumed behaviour, among other traits: *Homo europaeus*, *Homo asiaticus*, *Homo americanus* and *Homo africanus*. In Linnaeus' later editions of his seminal book *Systema Naturae*, he put these 'varieties', later translated as 'subspecies', into a hierarchy, inadvertently codifying the basis for the unscientific 'race science' that continues to persist.

While the Linnaean system attempted to categorize modern humans, it didn't define the genus 'Homo' or initially recognize the possibility that humans evolved from common ancestors shared with other members of the animal kingdom. It would be another thirty years before Charles Darwin published *On the Origin of Species*, and popularized the idea that natural selection could drive evolution.

The first unofficial Neanderthal remains – the partial cranium, jaw and some teeth of a two-year-old child – were found in 1829 in a group of caves along the banks of a stream just north of the Belgian municipality Engis; they were classified as modern humans, albeit old ones. Similarly, a skull discovered in a quarry in Gibraltar in 1848 was thought to belong to our species, and was stored in a cupboard for many years.

But when a collection of fossils was discovered in the Neander Valley in Germany in 1856, some scientists realized that there was something different about these bones. A few years later, they labelled the 40,000-year-old fossils – which included a skull-cap and several parts of the skeleton – the type specimen for *Homo neanderthalensis*, an archaic human ancestor. Initially, scholars argued that the find was actually a deformed or ill modern human, but as more fossils were found, the scientific community recognized *H. neanderthalensis* as a valid species. The Gibraltar and Belgian finds were eventually included in *Homo neanderthalensis*.

The discovery of an 'elderly' man (who would have been about forty years old when he died) at La Chapelle-aux-Saints in southern France in the early twentieth century fuelled the idea of Neanderthals as a degenerate, ape-like creature. The man suffered from severe osteoarthritis, which would have warped his bearing to some extent. However, a more recent study has suggested that the La Chapelle-aux-Saints man wouldn't have stooped quite as much as described,[36] which speaks to the describing scientist's own bias rather than the morphological reality.

The negative perception of Neanderthals – in which they were depicted as stupid, lumbering, cave-dwelling brutes – endured for many years, but we now acknowledge them as very like us: a cognitively capable people who thought, felt and engaged with the world around them.

An abundance of fossils and artefacts

We now know more about Neanderthals than we do about any other extinct human species. We have found thousands of fossils belonging to hundreds of individuals, including some nearly complete skeletons; we've managed to sequence their genomes; and we have unearthed many artefacts from their lives – from rock art and tools to possible musical instruments – as well as from their deaths, with possible intentional burial sites. Also, we have discovered the full spectrum of ages, from infants through to the elderly, which gives us a privileged perspective on this group of people.

But there is still a great deal we don't know. For example, we do not know how numerous Neanderthals were, although some estimates, based on their mitochondrial DNA, suggest that at their most abundant there were about 52,000 Neanderthals, before their numbers declined towards extinction. Their remains have been found from England to Siberia and from Uzbekistan to Palestine, showing that they roamed over large areas of Eurasia. Across such vast distances and geographic locations, Neanderthals would have had to cope with a variety of climates during their existence. They also endured for a relatively long time, from about 400,000 to 40,000 years ago, although we do not know what ultimately happened to them.

Ancient climates

The Earth has had at least five major ice ages, with the first happening about 2 billion years ago. We are currently in the most recent one, which began about 2.6 million years ago. In an ice age, the planet isn't in a constant frozen state – temperatures fluctuate between cold stretches (called glacials) and warmer times (called interglacial periods).

The most recent glacial period began about 115,000 years ago and ended 11,000 years ago. The average global temperature would have been about 7 degrees Celsius colder than today, meaning that large swathes of North America and Eurasia would have been covered in ice and, with more of the planet's water trapped in ice sheets and the polar ice caps, sea levels would have been much lower.

The ramifications of a colder climate would have extended beyond survival and physical comfort for bipedal hominins. With lower sea levels, Britain would have been connected by land to the rest of Europe, North Africa kissed Europe and the Middle East, there were land bridges between the islands of South East Asia, and Australia, Tasmania and Papua New Guinea formed a single supercontinent – and *Homo* individuals would have been able to walk to what would today be considered other continents.

Neanderthals had large skulls, with brains similar in size to those of modern humans. The middle of their face was quite big, with an ample prominent nose. Their front teeth were larger than ours, as was their jaw. But unlike Sapiens, they had no chins. They were stockier than us, an adaptation attributed to the colder environments in which they lived, and they seem to have been stronger.

It also appears that they may have had some form of language. There is only one bone associated with vocals: a dainty U-shaped bone called the hyoid. Despite the richness of the Neanderthal fossil record, only one hyoid has been discovered so far, and it was remarkably similar to that of humans. Researchers modelled what a Neanderthal voice box would have looked like, and found that it would have been a lot bigger than ours.[37] So while Neanderthals would have been able to make a fairly large range of sounds, they wouldn't have sounded quite like us. However, they listened to similar frequencies to us, hearing capabilities that would have evolved for a reason – and some researchers believe that that reason was communication.

Using high-resolution CT scans, researchers created 3D models of the ear structures of Neanderthals, modern humans and individuals from Sima de los Huesos.[38] The models mimic how air moves through the ear structure and how sound waves propagate into the inner ear. Neanderthal hearing appears to be particularly good at picking up the frequencies used in speaking, like humans – in particular for consonant sounds. However, the physical ability does not mean that they could actually communicate in language or that they had the same flair for language that *Homo sapiens* do. But it is possible.

A comparison of the Neanderthal and Sapiens genome in 2020 has yielded telling differences between our species.

By comparing modern humans, archaic humans and chimpanzees, scientists identified almost six hundred genes, which are expressed differently in modern humans.[39] The greatest changes relate to vocal and facial anatomy, and this trend is unique to modern humans. This suggests that Sapiens evolved to communicate both in language and facial expression.

However, we do not know when such mutations occurred after we split from our common ancestor. In fact, we are not certain when exactly Neanderthals split from the human lineage, and who our last common ancestor was. Scientists assume that *H. heidelbergensis* or possibly even *H. antecessor* was a common ancestor of Neanderthals and Sapiens.

A vast underground cave in the hills of Atapuerca, Spain contains some clues to this mystery and has rewritten part of our lineage's narrative. In a deep chasm, known as Sima de los Huesos (the 'Pit of the Bones'), researchers have uncovered more than 6,500 skeletal remains, belonging to about twenty-eight individuals. The hominin remains in the pit, which is at the bottom of a 13-metre shaft, have been dated to about 430,000 years old and mainly belonged to young adults and children.

They had brain sizes that overlapped with Neanderthals and Sapiens, most were right-handed, and they were more robust than modern humans. Scientists also suspect that the respective morphologies of the males and females were not that different, and so they lack sexual dimorphism which is similar to what is found in Sapiens. Some of the skulls show signs of trauma, which may have been their cause of death. The chasm has yielded only one tool – an Acheulean hand axe made of red stone, which has been called 'Excalibur'. Some scholars hypothesize that it was a ritual offering.

Molecular mystery

Initially, palaeoanthropologists thought these ill-fated hominins were *H. heidelbergensis*, but some suspected that their skull, the protruding middle face and their teeth suggested a closer affinity to Neanderthals. In 2013, new evidence confused the matter even further: scientists published a mitochondrial DNA sequence from one of Sima de los Huesos individuals, representing some of the oldest DNA ever recovered.[40]

The Neanderthal genome

In 2022, Swedish scientist Svante Pääbo won the Nobel Prize in Physiology or Medicine. In 2010, he had produced the first complete Neanderthal genome from 40,000-year-old bones. Numerous Neanderthal genomes have since been sequenced, and Pääbo is credited with establishing the field of ancient DNA studies. His research also showed that Sapiens and Neanderthals interbred, and Pääbo and his team retrieved DNA from a Denisovan finger bone.

Studies have shown that some modern humans have about 2 per cent Neanderthal DNA, and that makes certain modern populations susceptible to disease.

Strangely, we have so far found no Neanderthal mitochondrial DNA in modern humans.

The mtDNA showed that the individual was connected to the Denisovans, a mysterious group of extinct humans living in Asia at the same time as Neanderthals and modern Sapiens. However, scientists later managed to extract nuclear DNA from more Sima de los Huesos individuals, which showed that they were in fact early Neanderthals or very closely related.[41]

These molecular discoveries, and other recent finds, have shed light on the various evolutionary splits between the groups. First, the age of the Sima de los Huesos Neanderthals meant that these early or 'proto' Neanderthals would have been too close in time to known *H. heidelbergensis* for that group to be a direct ancestor. Consequently, a common ancestor between them must have been older, making *H. antecessor* – which was also found in Atapuerca – a possible candidate. Other scholars argue that humanity could still have evolved out of another older population of *H. heidelbergensis*.

Since the Sima de los Huesos Neanderthals contain both Neanderthal and Denisovan DNA, the split between the two groups must have occurred before 430,000 years ago. Moreover, an analysis of Neanderthals' and Sapiens' teeth morphology and evolution pushed back the date of divergence between the two to 800,000 years[42] – although not all scholars are convinced that they branched out this long ago. Some genetic studies put the split closer to 600,000 years ago, which coincides with the presence of *H. antecessor*.[43]

Another early Neanderthal was found in the English village of Swanscombe in Kent in the mid-1930s. Thought to be a female – based on fossilized skull fragments – this individual lived about 400,000 years ago. Many Acheulean tools were also unearthed there. Meanwhile, at a similarly aged site in the Essex town of Clacton-on-Sea in eastern England, in 1911 an amateur pre-historian discovered the oldest known wooden tool, the Clacton Spear. Made of yew, only the tip remains, but it is significantly older than the Schöningen Spears found in Germany.

We now know that Neanderthals were in Europe from 430,000 years ago, but our best-known examples of Neanderthals and their lives are from about 130,000 to 40,000 years ago, before they abruptly disappear from the fossil record.

During these long years, Neanderthals continued to evolve, and these changes can be tracked through fossil finds across Europe. 'Early' Neanderthals were taller, while 'Late' Neanderthals tended to be stockier, which some attribute to cold weather adaptations. Others suggest that their stockiness would have made them stronger.

Neanderthal technology and culture

The earliest popular impressions of Neanderthals were neither flattering nor accurate. The more we learn about them, the more they are shown to be people who were able to adapt to their environment, had complex social interactions, were excellent hunters, and were capable of technological innovation and art.

The species endured for more than 350,000 years, across a range of geographies, including freezing locations such as Denisova Cave in Siberia and the warmer woodlands of the Mediterranean coast. They would have had to adapt to the physical conditions and modify their hunting strategies as animal populations changed.

Did Neanderthals wear clothes?

Many of the places where Neanderthals lived were cold, and they survived numerous glacial periods. This has led palaeoanthropologists to suggest that they developed stockier builds to adapt to chilly temperatures. But some scholars suggest that they still would have had to cover up about 80 per cent of their bodies during winter.[44]

Since most clothing is made from perishable materials, such as skins, they don't usually withstand the ravages of time. However, in 2020 scientists described a 50,000-year-old fragment of cord made from the inner bark of a tree.[45] The three-ply string fragment was found at a site called Abri du Maras in France, which Neanderthals occupied numerous times.

Additionally, human head lice and clothing lice diverged between 170,000 and 83,000 years ago, according to a 2011 study.[46] Researchers assume that around this time humans began wearing clothes, as some hair-living lice moved to a new habitat: clothing. But even if someone was wearing clothes, we don't know who: a number of *Homo* species were around at that time.

An important factor in Neanderthals' resilience was their technology. While older hominins used pebble tools (Oldowan technology or Mode 1) and pear-shaped hand axes (Acheulean technology or Mode 2), Neanderthals are associated with Mousterian technology, known as Mode 3. From about 300,000 to 35,000 years ago, we see a decrease in Acheulean tools and an increase in this more modern innovation.

Named after a site in the Dordogne region of France, known as Le Moustier, these artefacts were often formed from a pre-prepared core, and humans used them for a variety of tasks, including butchery and woodworking. The major innovation that enabled the creation of Mousterian tools is a technique called 'Levallois technique'.

It was named after the Levallois-Perret suburb of Paris, France, where examples of it were found in the nineteenth century. The method involves chipping pieces of stone ('knapping') off a 'core' rock using another smaller rock. After creating the desired shape of the flake, a single blow separates the flake from the core rock. This approach allows the tool-maker much more control as they can create the final shape of the flake before striking it off. It is surprisingly tricky to do, and requires insight into fracture weaknesses and different types of stone.

There are examples of Levallois technology as early as 300,000 years ago, but it really comes into its own in European Neanderthals' Mousterian tools from roughly 160,000 to 40,000 years ago. From the existence of such items, we can infer a number of things about their creators: they were able to plan and choose the right core material, and they had strong hands that could grip and control their tool-making stones. Some studies have tracked the geographic distances between where implements were found and where their makers would have sourced the raw materials, showing the lifestyles that different groups of Neanderthals had. Some communities sourced their tools from nearby geological resources, while others travelled long distances to get them.

Researchers have also shown that Neanderthals used resin to glue their stone instruments together, something which would have required the use of controlled fire. Also, seashell tools in Italy indicate that those Neanderthals were able to dive into water, maybe as deep as 4 metres, to collect clams whose shells they fashioned into implements.[47] Interestingly, Neanderthals seem to have shared their technology know-how with each other, and there are many debates around whether the group copied Sapiens' technology or developed more advanced tool-making techniques on their own as time passed.

What did Neanderthals eat?

Evidence from numerous Neanderthal sites shows that they ate meat, with some scholars suggesting that they preferred to hunt adult animals rather than younger ones (and thus preserving population sustainability). Carbon and nitrogen stable isotope ratios in their bone collagen have shown that they were top predators, eating large herbivores, with a diet that remained fairly similar between 120,000 and 37,000 years ago throughout Europe. Their bodies often bore the distinctive signs of trauma, similar to those of professional rodeo riders, suggesting that it was common for Neanderthals to have hostile encounters with large, aggressive animals. They even sometimes ate each other: the butchery marks on Neanderthal bones in Moula-Guercy, France, resemble the marks on reindeer from the same site.

They were also resilient and it appears that their diet depended on what was available. In Gruta da Figueira Brava, a cave in the Arrábida Nature Reserve in Portugal which Neanderthals inhabited about 100,000 years ago, researchers have found more than 560 fish bones, as well as the remains of seabirds and shellfish. Prior to that study, many palaeoanthropologists thought that fishing was the province of Sapiens.[48]

A 2021 study of tooth bacteria rejigged our ideas about what our *Homo* cousins and ancestors ate.[49] The researchers analysed the bacterial communities covering the teeth of Neanderthals and modern humans, and compared them with those of chimpanzees, gorillas and howler monkeys. A major finding was that both Sapiens and Neanderthals had streptococci bacteria in their mouths, which is required to digest starches. The authors suggest that the two *Homo* species inherited the microbes from their common ancestor, who lived between 800,000 and 600,000 years ago. In a reply to this study, though, other scientists questioned their assumptions,[50] and the two camps continue to debate the issue.[51]

What Neanderthals ate also depends on whether they were using fire to prepare it. We know that they did use fire, particularly later Neanderthals, but it is still unclear whether they started their own fires or foraged flames from elsewhere. If they foraged it, they would have been reliant on finding it first – and so their cooked food would have been opportunistic rather than habitual.

Evidence of complex behaviour and culture

We do not know the extent of Neanderthals' cognitive capabilities, and this area of research is subject to robust debate with new evidence both in favour of and against the heavy-browed species' intellectual abilities. Symbolic thinking is a hallmark of *Homo sapiens*, a way of approaching the world that allows for abstraction, imagination and communicating complex thoughts. We do not know if Neanderthals could think in this way.

One of the major difficulties is being certain that Neanderthals were in fact the individuals who performed an act or produced an artefact. By the time we start seeing complex behaviour in the archaeological record, Sapiens may have been present – although we don't know when and where the two may have interacted – so we often can't say definitively that Neanderthals were responsible. And even if they were responsible, did they spontaneously develop their behaviour themselves or was it learnt from Sapiens or possibly even other *Homo* species?

However, perhaps the main problem is ascribing intent and superimposing our ideas of behaviour onto a species that disappeared 40,000 years ago. Additionally, it is not always possible to say definitively what an object was used for.

Nevertheless, there is some evidence that Neanderthals were capable of complex, artistic behaviour. They appear to have made jewellery, crafting beads of animal teeth and talons, shells and ivory. Eight ancient eagle talons, from at least three different birds, in a rock shelter near Krapina in northern Croatia suggest that 130,000 years ago Neanderthals may have been fashioning them into jewellery. Extracted proteins from bone fragments found in a cave at Arcy-sur-Cure in France proved that the bones, found alongside jewellery, did in fact belong to Neanderthals who died about 42,000 years ago.

Meanwhile, a flute, made from a cave bear's thighbone, could possibly be humanity's oldest known instrument. Unearthed in a cave near Cerkno in western Slovenia in 1995, it is 11 centimetres long and has been dated to between 60,000 and 50,000 years old. In some places, it is called the 'Neanderthal flute'. The piece of bone has two complete holes in it, and many pages of argument have been dedicated to whether these holes were intentionally formed by clever hands or a carnivore's teeth. If it was indeed carved by humans, they would have been in Europe before we think modern Sapiens entered the scene. This has led some scholars to believe the 'flute' is proof that Neanderthals were capable of making music. However, it should be noted that even if Neanderthals made the 'flute', there is a big difference between making sounds to communicate with other people (such as a warning or other signals) and composing music. Other bones with holes in them were discovered in the 1920s and 1930s at other sites in the mountains of Slovenia, but most were destroyed in an Allied air raid during the Second World War.

The red ladder symbol, painted on the La Pasiega cave in Spain, has a minimum age of 64,000 years. It is thought to have been painted by a Neanderthal.

Some argue that the existence of rock art is a sign of complex, possibly symbolic, thought, but rock art is notoriously difficult to date – and to ascribe to a specific group of humans. Researchers have uncovered ochre-stained shells with holes in them, red and yellow pigments, and shell containers with the telltale signs of pigment mixtures at Cueva de los Aviones in Spain. These 'painting tools' have been dated to about 120,000 years old. A few years ago, we thought that this was much earlier than the arrival of Sapiens, but new dates for Sapiens fossils from the Apidima Cave in southern Greece suggest that Sapiens were adventuring in Europe about 210,000 years ago. So, the 'tools' could have been made by Neanderthals or Sapiens, although the balance of evidence currently suggests they were made by Neanderthals.

Similarly in Spain, scientists have dated the thin mineral crusts covering the rock art at three different sites using uranium-thorium dating. (Mineral deposits can accumulate on rocks, and multiple layers can build up over time, for example on stalagmites and stalactites in caves. Over time, such deposits can collect in a fine film over rock art.) In Spain, red dots and lines stand out against the brown and white rock, despite the thin crust.

The researchers dated the films coating the art to about 64,000 years ago, meaning that the art underneath was a similar age or older, and could have been made by Neanderthals or Sapiens, although most scientists think it was created by Sapiens.

Meanwhile on Sulawesi island in Indonesia, archaeologists discovered an example of figurative painting dated to at least 43,900 years old. Discovered in a cave called Leang Tedongnge in December 2017, scientists published a description of the painting in 2021.[52] The 4.5-metre-wide panel in a limestone cave depicts several human-like figures hunting animals, including Sulawesi pigs. Interestingly, the animals seem to have different ages, ranging from 41,000 to 43,900 years ago. This panel is thought to be humanity's oldest recorded story and figurative painting, depicting the lives of these people. However, we do not know who the artists were: whether Sapiens made it to South East Asia earlier than we thought or if another group was responsible, since we believe that there were a number of species in the region at that time.

Prior to these recent discoveries, a painted rhino in the Chauvet Cave in south-eastern France was thought to be the oldest figurative painting. It remains the oldest in Europe, at between 35,000 and 39,000 years. But once again, its authorship is ambiguous: Neanderthals were not the only humans around at that time, and it could have possibly been early Sapiens.

Another marker of culture and complexity is social interaction. A particularly fraught topic in palaeoanthropology is whether Neanderthals buried their dead – a trait associated with the ritualistic, modern behaviour of Sapiens. It speaks to a higher level of cognition than we currently ascribe to Neanderthals. Another possibility is that, if Neanderthals did behave in this way, they learnt such behaviours from Sapiens. It is important to note that intentional burial, with funeral activity, is different from finding a convenient place to dump bodies. There are also numerous examples of cultural exchange flowing the other way, with Neanderthals sharing their practices with Sapiens, such as the use of eagle talons in decoration.[53]

In 1908, prehistorians discovered the remains of an elderly Neanderthal at La Chapelle-aux-Saints in France. In the years that followed, researchers found more individuals at the site, leading anthropologists to ponder the matter of ancient human burial. This debate raged for many years. Scientists re-examined the cave and excavated it further, and determined that the Neanderthal remains had been deposited there on purpose. (That does not necessarily mean they were buried, or were given any form of funeral.) Incidentally, the 'Old Man' of La Chapelle-aux-Saints, having lost most of his teeth during his life, would have had to rely on other members of the community to assist him, or would have received preferential treatment. His existence suggests social care and support among Neanderthals.

In 2020, scientists argued that someone intentionally buried a two-year-old Neanderthal child in the Dordogne region of France almost 42,000 years ago.[54] Initially excavated in the late 1960s and early 1970s, the site was re-examined by palaeoanthropologists in 2014 and it was decided that someone had intentionally dug a pit to deposit the child's body into, before covering it with the older sediments.

Further west in Iraqi Kurdistan, researchers have said that the individuals discovered there were buried with formal funeral rites, including flowers. Between 1951 and 1960, archaeologists uncovered the remains of ten Neanderthal men, women and children in the Shanidar Cave, which were later estimated to be between 65,000 and 35,000 years old. Some of them are thought to have died due to a rockfall, but one of the bodies was associated with clumps of pollen grains. These pollen grains, the researchers argue, suggest that flowers were intentionally placed with the corpse. Other palaeoanthropologists have dismissed the 'pollen burial' hypothesis, saying that pollen could have been brought in by insects or even flower-loving rats.

The debate about funeral behaviour is far from decided.

It continues to rage because such behaviour indicates a level of cognition comparable to ours. Many of the traits that we thought made us special – bipedalism, giant brains, tool technology and social networks – were actually present long before Sapiens appeared. Funeral and burial rites, however, are still possibly the sole province of Sapiens.

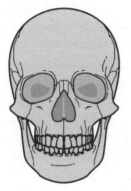 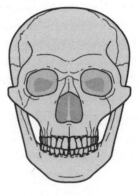

Sapiens versus Neanderthal skull

Denisovans

In Siberia, Russia, there is a cave alongside a river. It is about the size of a modern four-bedroom home, and in the 1700s it was, legend has it, inhabited by a religious hermit called Denis. The hominin fossils in the cave, known as Denisova Cave after the hermit, have overhauled our understanding of other *Homo* species and how they related to us Sapiens.

The cave's three chambers have yielded a wealth of animal fossils, bone artefacts and stone tools, dated to between

125,000 and 180,000 years ago. They also contained more recent jewellery made from sculpted bones and the teeth of deer.

But in 2008, Russian archaeologists uncovered a hominin finger bone. They have since found numerous hominin bone fragments, including part of a toe. The cold temperatures in the cave preserved some of the DNA material in some of the bones.

By analysing DNA in the finger bone, which is thought to have belonged to a young woman, scientists produced the full genome of this unknown human species, which have been labelled Denisovans. She was genetically distinct from both Neanderthals and modern humans, and lived between 76,200 and 51,600 years ago. There are also older Denisovan remains, dating from about 217,000 years ago, although some of the artefacts in the cave are much more ancient – about 287,000 years old – indicating that the place had been inhabited before these individuals died.

Her DNA suggests that Denisovans split off from Neanderthals between 300,000 and 400,000 years ago, and that Neanderthals and Denisovans were more closely related to each other than to modern humans. Neanderthals and Sapiens split from their last common ancestor about 800,000 years ago. Denisovans are thought to have lived across Asia. A mandible from the Baishiya Karst Cave on the Tibetan Plateau could have belonged to a Denisovan, and in 2020 scientists reported that they found Denisovan mitochondrial DNA from sediments that are between 100,000 and 60,000 years ago.[55] There are only a handful – literally – of Denisovan fossils, so we have no idea what they looked like and whether they resembled modern humans or Neanderthals.

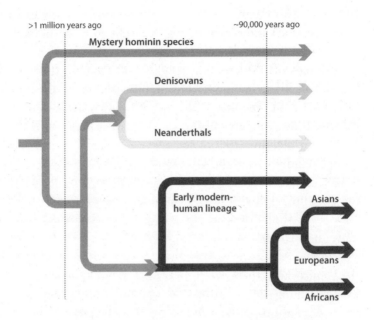

>1 million years ago   ~90,000 years ago

Mystery hominin species

Denisovans

Neanderthals

**Early modern-
human lineage**   Asians

                                         Europeans

                                         Africans

Scientists think that Denisovans split off from
Neanderthals between 300,000 and 400,000 years ago.

The toe bone at Denisova Cave, however, belonged to a
Neanderthal, and was in the same layer as the Denisovan
finger bone. Known as the Altai Neanderthal, this individual
is thought to have lived about 120,000 years ago.

But perhaps the most astounding discovery to have come
out of the Denisova Cave was a long bone fragment from a
young female who died about 90,000 years ago, which the
scientists have nicknamed Denny. Denny's mother was a
Neanderthal and her father was a Denisovan. This is the latest
example of *Homo* interbreeding to have been identified in the

last decade, although it is one of the most surprising. Twenty years ago, such interbreeding was thought to be so unlikely that it was impossible.

Analyses of modern human DNA show that there are individuals living in Asia today who have traces of Denisovan DNA in their genome. One of these Denisovan genes, EPAS1, allows modern Tibetans to survive at high altitudes.

Meanwhile, in 2019, scientists described a Denisovan partial mandible from the Tibetan plateau in China (although it had in fact been discovered years earlier).[56] Collagen in the mandible linked it to the individuals from the Denisova Cave, even though scientists dated the Chinese specimen to more than 160,000 years old.

Other fossils from China could also possibly belong to the Denisovans. These hominin fossils, found at a number of localities, range in age from 350,000 to 100,000 years ago, and palaeoanthropologists have speculated whether they fall into *H. erectus*, *H. heidelbergensis* or Neanderthals.

One of these specimens is Dali Man. In 1978, a farmer in Dali County, in the Chinese province of Shaanxi, discovered a complete fossilized skull. It appeared to belong to a species that sat between Peking Man (*H. erectus*) and Neanderthals.

It had a long, low skull and an extremely low forehead, with a brain size of just over 1,100 cubic centimetres (which is at the bottom end of Sapiens' brain size). But it also had a prominent sagittal keel, the ridge of bone seen in *H. erectus* and which is strongly pronounced in the likes of *Paranthropus boisei*, with its strong chew capabilities. It also had massive brows, but unlike Peking Man, whose brows formed a long imposing ridge across his face, Dali Man's brows arched over each eye. Initially, it was labelled *Homo sapiens daliensis*, although it has sometimes been referred to as *Homo heidelbergensis daliensis* (as the Out of Africa 1 hypothesis gained traction), and often just 'archaic *H. sapiens*'. Since the discovery of the Denisovans, some scientists have suggested that Dali Man may in fact have links to these Siberian hominins, but we need more evidence to substantiate such a relationship.

In 2021, Chinese palaeoscientists described a fossilized skull that was about 140,000 years old.[57] It had actually been discovered by a labourer eighty-five years earlier and hidden in an abandoned well (just four years earlier, researchers had discovered Peking Man near Beijing). The scientists called the cranium *Homo longi*, derived from the Long Jiang, or Dragon River, in the Heilongjiang province of China. Its nickname is Dragon Man.

The skull belonged to a male who, like Dali Man, had massive arching brow ridges. He had a flat face, a rather large, round nose, and a brain capacity of about 1,420 cubic centimetres – similar to that of modern humans and Neanderthals. Given the skull's relatively modern appearance, its discoverers suggest that it could have been a close human ancestor, possibly closer than Neanderthals. Some say that it could have been a Denisovan, and yet others think it should be placed with *H. daliensis*. But without DNA evidence, such theories we cannot truly know. For now, DNA evidence points to Sapiens originating in Africa.

The secrets of collagen

In 2009, scientists at the University of York first described a technique called zooarchaeology by mass spectroscopy (or ZooMS). They initially developed it to distinguish between a sheep and a goat, based on the extraction of collagen from an animal's bones. Scientists can break down the collagen into its constituent peptides, which are chains of amino acids, and identify the 'fingerprints' that identify different animals. The method has since been refined extensively. Denny was the first hominin this technique has been used on, and with it, scientists identified her bone as belonging to a hominin, rather than one of the other animals found at the site. The technique has since been used to determine other hominin specimens at the site.

PART 4

THE RISE OF THE 'WISE MAN'

10 THE FIRST SAPIENS

Today, there is only one human species on the planet: *Homo sapiens*. We know that anatomically modern humans emerged about 300,000 years ago, and then proceeded to dominate the world, becoming the most common species of primate. Despite how different we may look, all humans living today belong to Sapiens. But the question of where and from whom our species evolved continues to be the source of much debate – as it has been since we first recognized that our species did not arrive on Earth fully formed.

When Carl Linnaeus named our species *H. sapiens* (meaning 'wise man') in the 1700s, it would have been comical – or, more accurately, heretical – to suggest that humans had in fact shared a common ancestor with chimpanzees between 5 and 9 million years ago. But as people unearthed ancient hominin remains with increasing frequency, the narrative of our origins began to shift. And the more we learn and discover about our evolution, the more complicated and complex that narrative becomes.

Scientists first began to realize the antiquity of our species in the late nineteenth century. In 1868, railway construction near the French village of Les Eyzies the Dordogne revealed a rock shelter containing the remains of four adult humans and an infant. They were named Cro-Magnons (which later became synonymous with anatomically modern humans in Europe before falling out of use) and appeared to have been intentionally buried in the rock shelter, along with tools and fossilized animal bones, some of which may have been fashioned into jewellery. The associated artefacts have been dated to between 32,000 and 30,000 years old.

These Sapiens differed from the *Homo* species that we now know came before us. They had tall, rounded skulls with an almost vertical forehead. They didn't have the startling frowning brow of Neanderthals and *H. erectus*, and nor did they have the protruding face and jaw of some of our older relatives. Notably, they had chins, which is something no other *Homo* species had.

But they are not exactly the same as Sapiens today. For one thing, they had larger brains – between 100 and 200 cubic centimetres larger than our average 1,350 cubic centimetres. Their bones also tended to be thicker. The remains of these first Cro-Magnons showed that they lived a hard life: they had broken bones and survived traumatic injuries. The face of one of the men was covered with pitted lesions, which for years was thought to have been caused by a ravaging fungal infection. In 2018, researchers found that he may have had neurofibromatosis, which would have caused tumours on his face and possibly even in his ear. Practically, what this suggests is that these people would have had to look after each other, or would have been cared for, during the course of their lives, indicating support and attention.

Initially, the discovery of ancient humans ruffled many scientific and societal feathers, and scientists interpreted them from within the mainstream nineteenth-century religious creationist model. Commentators suggested that these ancient humans were 'antediluvian' people who had been wiped out by the biblical Great Flood, and some even said the bones and associated artefacts pointed to witchcraft. However, they did confirm one prevailing bias: humans originated in Europe. This idea was reinforced by the fraudulent Piltdown Man in England, and it was many years before people believed the evidence that pointed to Africa as the real birthplace of not only hominins but also Sapiens.

Early humans

Today, it is widely accepted that *Homo sapiens* first appeared on the African continent, based on the fossil evidence and genetics of modern humans, but it is uncertain where. A 2023 study exploded many of our certainties around the birthplace of Sapiens: by analysing the genomes of 290 living people, the researchers concluded that we did not suddenly appear in one place at a particular moment in history. Instead, modern humans descended from at least two populations that lived on the continent for about a million years, before merging in numerous independent interactions.

Another issue of contention is whether Sapiens evolved its 'modern' form rapidly about 200,000 years ago, or gradually over the last 400,000 years. While early Sapiens are included within the species, they did not look precisely as we do today.

So far, the oldest evidence we have is from the dusty, arid landscape of Jebel Irhoud in Morocco. Researchers have known about the old limestone cave site there since the early 1960s, having discovered stone tools – which they attributed to Neanderthals – and human remains, which they thought were fairly recent (archaeologically speaking). But new excavations yielded human remains belonging to five individuals: three adults, a juvenile and an eight-year-old child. They resembled modern humans more than any other *Homo* species.

Their faces were tucked into the bottom of the skull and they had relatively small brows, but they had comparatively large teeth and an elongated brain case. When scientists re-evaluated the site's age, using thermoluminescence dating on a tooth and stone tools associated with the remains, they found that the humans were about 315,000 years old. This age is significantly more ancient than the previous oldest Sapiens remains. Interestingly, the flint tools were also not made of local stone, but came from a region about 50 kilometres south of Jebel Irhoud, showing that these people had brought their tools with them. The implements also bear the telltale traces of fire.

The Jebel Irhoud humans threw the prevailing narrative into disarray – even before the 2023 pan-African evidence. Many scientists had thought that eastern Africa was the birthplace of Sapiens. The idea that Sapiens arose from multiple populations in Africa had been gaining momentum over the years, but in 2023, scientists offered firm evidence backing it up. Before the Jebel Irhoud dating, the oldest Sapiens had been in Omo Kibish, Ethiopia, thousands of kilometres to the south-east. Like the Jebel Irhoud fossils, some of the human remains (including two partial skulls) at Omo Kibish were discovered in the 1960s. The two skulls were found a few kilometres apart, and one is more modern-looking than the other, which has some archaic traits. New dating, using more modern techniques, has estimated that these people died about 230,000 years ago, much later than initially thought.[58] The diversity in the two skulls raises some confounding questions about whether they represent noticeable variation within the same population, or if two groups of different-looking Sapiens lived in close proximity to each other.

Similarly, a skull unearthed in Florisbad in South Africa, even further away from Morocco than Ethiopia, continues to skew any easy and simple narrative of Sapiens. Found in the 1930s, this individual would have been as large as a modern human, with a greater brain volume of about 1,400 cubic centimetres. In 1996, using electron-spin-resonance dating on tooth enamel, scientists dated the skull to about 260,000 years (give or take 35,000 years). South Africa is also home to some of the earliest examples of complex Sapiens culture, which – along with modern genetic evidence – has led some scholars to argue for southern Africa as the origin of Sapiens. In 2023, scientists announced the discovery of the oldest known Sapiens footprints – dated to about 153,000 years old – captured in rock on South Africa's southern Cape coast.

What all these discoveries mean is that there were Sapiens, with an array of modern human features and some archaic ones, living very far away from each other on the same continent. Considering the new genetic evidence for a pan-African Sapiens origin, these discoveries now make much more sense – and another piece of the human evolution puzzle falls into place. More fossil discoveries in the future, or redating of old ones, will no doubt both simplify and complicate the narrative of our origins even further.

Who did we evolve from?

In short, we do not know definitively which species we evolved from. The matter of our last common ancestor, and who we descended from, is still an open question. Some argue that big-brained *Homo heidelbergensis* offers the best possible candidate; and examples of this species have been discovered at numerous sites in Africa. It is possible that *H. heidelbergensis* displaced *Homo ergaster* (or African *Homo erectus*).

Other scholars believe that *Homo antecessor* is a better contender. We know about them from a site in the Atapuerca hills in Spain, and they look surprisingly modern (even though we have not found a complete skull).

In 2019, scientists used computer modelling to compute what the last common ancestor of all modern humans would have looked like.[59] They compared early human fossils from locations such as Florisbad, Jebel Irhoud and Omo Kibish, looking at their structure, how they related to each other, and whether there were evolutionary links between their morphology. To make sure that they made allowances for intra-species variation, they also repeated the process for modern Sapiens, including populations from a range of places, as well as extinct humans such as Neanderthals. The resultant early Sapiens skull looked a lot like us, but not quite. Notably, it had more pronounced brow ridges and its lower face stuck out more than ours do.

The modellers hypothesized that our evolutionary lineage actually produced a number of populations in Africa around 350,000 years ago, and that populations in eastern and South Africa interbred to produce us. At the time, other scientists cautioned that new fossil finds may once again skew the data, and thus the researchers' findings. The genetic evidence for a pan-African origin seems to bolster the model, but new findings will no doubt enrich what we now think we know.

Out of Africa: take two (or three or four)?

Back in the early-to-mid-twentieth century, scientists mainly relied on physical fossils and prehistoric artefacts to write the story of humanity, but technological developments have given us more tools to nail down plot points. In terms of dating, new techniques and more accurate methods have allowed researchers to pinpoint when hominins lived, but the revolution in and ubiquity of DNA sequences have really illuminated the story of our origins. However, these two branches of evidence – the morphological fossil data and DNA information – can sometimes pull the narrative in different directions and leave us with more questions than answers.

Homo sapiens are distinct from the human species that came before us, in terms of both our skeletal morphology and the cultural artefacts that we have left behind. Physically, we have distinctly high, rounded skulls, and a flat vertical face, as opposed to the protruding faces of our ancestors. We have also lost the sagittal crest of earlier hominins, which once anchored strong chewing muscles. We have small front teeth, and – unique among *Homo* and other hominins – a protruding chin, even in infancy. Our limbs are long and thin, a marked departure from the more robust likes of *Paranthropus boisei* or even Neanderthals, and our short, narrow pelvis, with a large surface in the hip joint, makes us excellent walkers and runners. Meanwhile, our hands sport long fingertips, capable of strong and precise grips.

Once again, palaeoanthropologists disagree about which

fossils to include within Sapiens. The earliest possible Sapiens do not look like modern-day humans; they still retained some archaic features – and not necessarily the same ones across different localities. Some argue that only individuals who looked like living humans should be included under the umbrella of *H. sapiens*. Others prefer to call early incarnations of Sapiens 'ancient *H. sapiens*', but still group them under the species' umbrella. Even more hard-line lumpers think that there should only be two species within *Homo*: *H. erectus* and *H. sapiens*.

The oldest possible *Homo sapiens* we know of lived in modern-day Morocco about 315,000 years ago, and then, many years later, Sapiens' remains began cropping up in other parts of the world.

Current thinking, based on the fossil evidence, is that Sapiens left Africa in a series of 'pulses', or waves of migration, most of which were unsuccessful in creating permanent settlements. They may have travelled through the Nile Valley and into modern Israel, where several Sapiens remains have been found, or across the Bab al-Mandab strait (also known as the 'Gate of Grief') linking the Horn of Africa to the Arabian Peninsula when planetary bouts of cold meant the sea level was lower.

The oldest known example of Sapiens outside of Africa are in Greece – a relatively recent finding that left many scientists flummoxed. In 1978, Greek anthropologists discovered two skulls in a block of breccia wedged high into the walls of the Apidima Cave in southern Greece. Apidima 1 and Apidima 2, as the partial crania were named, resembled modern humans and Neanderthals respectively, but dating had been inconclusive. In 2019, scientists managed to put rather remarkable dates onto the skulls: around 210,000 and 170,000 years respectively.[60] The researchers have dismissed the idea that round-skulled Apidima 1 was an early Neanderthal or similar to those found at Sima de los Huesos. It has been included in *Homo sapiens*, which means that modern humans were walking the landscape of Europe long before we thought that they did. There is some evidence that Neanderthals may have interbred with these early Sapiens explorers.

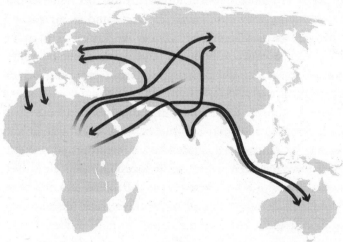

Scientists suspect that hominins have made numerous forays out of Africa, even possibly Australopithecines. Complicating the narrative of human evolution and migration, it is also possible that species returned to the continent.

Israel has also yielded a treasure trove of archaic Sapiens remains, indicating an ancient migratory route across the Levant. In 2018, scientists announced that a fragment of a Sapiens' upper jaw discovered in Misliya Cave at Mount

Carmel was between 177,000 and 194,000 years old.[61] In the twentieth century, early human fossils were discovered in the Skhul and Qafzeh Caves, also in Israel, but dating has been tricky, and they are assumed to be between 80,000 and 120,000 years old. At Qafzeh, the remains of up to fifteen individuals were found with seventy-one pieces of red ochre and ochre-stained tools, which have led some researchers to suggest that this is humanity's first recorded intentional burial.

Aside from the Apidima 1 skull in Greece, the earliest known presence of modern humans in Europe is between 51,700 and 56,800 years ago at Grotte Mandrin in south-eastern France. This would have put them in the same region concurrently with Neanderthals. Prior to that, we thought Sapiens had only settled in Europe between 45,000 and 43,000 years ago, thanks to teeth at three sites in Italy and a fourth in Bulgaria. The last known Neanderthal remains in Europe have been dated to, at oldest, 42,000 years old, showing that there was overlap between the species.

And this is where DNA disrupts many aspects of our origins story.

Mitochondrial Eve

In 1987, three scientists from the University of California, Berkeley in the United States published a remarkable study.[62] They analysed the mitochondrial DNA from 147 people from five geographic locations. They found that all of their DNA stemmed from one woman who is thought to have lived 200,000 years ago, 'probably in Africa'.

This paper, much to the scientists' dismay, gave rise to the name of a 'Mitochondrial Eve', with its biblical and unscientific connotations of Adam and Eve. What the science actually showed was that modern humans were all descended from a single woman who lived about 200,000 years ago.

What is a haplogroup?

A haplogroup is a group of people who genetically share a common ancestor. Haplogroups are assigned letters, with the L-type almost always consisting of populations from Africa, while Q refers to populations such as Melanesian, Polynesian and New Guinean. L0, one of the two most ancient clades, is associated with people ancestral to southern Africa's first people, the Khoisan. The other clade, which contains L1, L2, L3, L4, L5 and L6, split off from L0 about 170,000 years ago. L3 is the haplogroup most closely associated with the 'Recent Out of Africa' event, in which Sapiens migrated out of Africa about 70,000 years ago and journeyed across the planet.

While there were some problems with the initial study, follow-up research broadly confirmed their findings. Later studies narrowed down their findings, showing that Sapiens had migrated out of Africa in an important foray between 70,000 and 60,000 years ago (known as a 'Recent Out of Africa' event). The research also confirmed that Mitochondrial Eve had lived in Africa. People in southern Africa have the greatest genetic diversity of modern humans, consistent with an African origin of modern humans. As groups of humans migrated away from Africa, their genetic diversity declined. This is known as the 'Serial Founder Effect': if a population moves far away from its original larger group, they carry only a fraction of the genetic diversity of the whole group. If they split further, their genetic diversity continues to decline.

Mitochondrial DNA analyses show that by about 130,000 years ago, there were two distinct groups of anatomically modern humans in Africa: one in southern Africa and another in Central and eastern Africa. Such a migration would have coincided with a 'mega drought' in the region between 135,000 and 75,000 years ago. The drought is thought to have ultimately triggered a large (or, at least, successful) Sapiens' migration out of Africa about 60,000 years ago.

A fascinating 2007 study tracked the genetic diversity of *Helicobacter pylori*, a common infectious bacterium that can cause stomach ulcers, and found that the genetic diversity of *H. pylori* also decreases with geographic distance from Africa.[63] The study's authors posit that *H. pylori* seems to have spread from eastern Africa about 58,000 years ago, and that infected migrating humans took the microbe with them.

Before the seminal 1987 study, many scholars believed in a 'multi-regional hypothesis' for human evolution. The 1987 study, and the research that came later, has swayed scientific consensus toward a Recent Out of Africa model. However, recent DNA discoveries regarding the presence of Neanderthal and Denisovan DNA in modern humans suggests that the narrative may not be as simple as the Recent Out of Africa model suggests. There is evidence that Sapiens made a number of forays out of Africa – some as early as 200,000 years ago – as opposed to one single migration 60,000 years ago. It's also possible that these early Sapiens bred with humans they encountered on their travels and added to their genetic ancestry.

Multi-regional model

The strongest form of this theory held that Sapiens evolved from archaic humans, such as Neanderthals and *H. erectus*, simultaneously to create diverse populations around the world. Neanderthals evolved into Europeans, *Homo erectus* became modern Asian populations, and so forth. Current incarnations of this argument are more nuanced, and suggest admixtures of different human populations.

But new discoveries have shown that, as we have come to expect from palaeoanthropology, our story is not that simple. In 2010, Svante Pääbo and colleagues sequenced a Neanderthal genome and showed that Sapiens had sex with Neanderthals and had children. The DNA fingerprints of these assignations were written into modern human DNA, with populations in Europe and Asia having up to 2 per cent Neanderthal DNA. Later that year, Pääbo and collaborators extracted and sequenced DNA from a finger bone shard found in the Denisova Cave in Siberia. By comparing those DNA sequences with living people, they discovered traces of Denisovan DNA in modern human populations, namely

East Asians and Native Americans. Some groups in South East Asia, Oceania and Australasia have as much as 4 per cent Denisovan DNA. Denisovans and Neanderthals had definitely split by 430,000 years ago, since one of the humans at Sima de los Huesos contained distinct DNA from both; scientists suspect that they separated about 600,000 years ago, possibly in Africa.

These slivers of DNA can have important implications for our health. For example, Pääbo and Hugo Zeberg, a geneticist at the Karolinska Institute in Sweden, found that people with Neanderthal DNA were more likely to get a severe form of Covid-19 from the SARS-CoV-2 virus. Meanwhile, a Denisovan gene allows Tibetans to survive at such high altitudes.

But perhaps one of the most fascinating doors that ancient DNA opened is the matter of ghost humans, whose DNA is part of our genome but whose physical remains we have not found. (Or, if we have found them, we have not dated and recognized them for what they are.) In 2020, two researchers from the University of California, Los Angeles discovered that some people's DNA contained portions of code from an extinct branch of humans, which split from the Sapiens lineage less than 1 million years ago.[64] The geneticists obtained the genomes of more than 400 people currently living in West Africa, namely populations in modern-day Nigeria, Sierra Leone and the Gambia. They investigated how new gene variants arose in each population, and their data-crunching showed that part of these people's genetic ancestry – between 2 and 19 per cent – came from an ancestor that did not give its genetic material to other Sapiens populations around the world. They estimate that the ancient humans interbred with Sapiens in West Africa at some point in the last 124,000 years.

We don't know who this 'ghost population' was, and palaeoanthropologists will be combing through existing fossils, and hunting for new ones, to identify this 'ghost' ancestor.

Ancient DNA 'gold rush'

In 2023, the tally of sequenced ancient-human genomes surpassed 10,000. The first genome was sequenced in 2010, and belonged to a man who lived in Greenland about 4,000 years ago. Of the 10,000 genomes, more than two-thirds are from Europe and Russia, with African genomes only accounting for 3 per cent of the total. Part of the reason for the inequity is the wealth of recent European samples from about 12,000 years ago, when the last ice age ended. These samples also tend to have higher quality DNA because they are more recent.

The spread out of Africa

Homo sapiens were not the first hominins to make the journey out of Africa. We know that *Homo ergaster*, at least, left the continent and made it as far as Asia, where we know it as *Homo erectus*. However, scientists suspect that hominins may have made several forays out of Africa during our evolution, probably linked to ecological change, climate variability and the opportunities offered by lower sea levels. People may even have returned to the continent after adventures abroad.

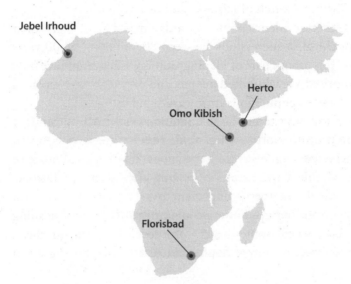

Rather than a single 'Garden of Eden' origin
for Sapiens, scientists now believe that our species arose
in several places on the African continent and interbred.

There are three ways that scientists track *Homo sapiens'* migration across the globe: their DNA, fossilized remains and the artefacts they left behind. It is a complex subject, and privy to several strands of baffling and sometimes contradictory evidence. Sapiens appear to have made a number of treks, known as 'pulses', out of Africa. This would explain some of our perplexing fossil finds, such as the 210,000-year-old Apidima 1 Sapiens skull in Greece. But between 135,000 and 75,000 years ago, there was a 'mega drought' in tropical

Africa. The water levels in Lake Malawi, one of the deepest lakes in the world, dropped by at least 600 metres. Scientists argue that such drastic ecological change pushed Sapiens out of Africa, in search of less arid and brutal conditions.

Genetic evidence points towards a major migration – or at least the most successful – between 80,000 and 60,000 years ago. This group, part of Haplogroup L3, is likely to have only comprised a few hundred people, and they may have replaced the Sapiens populations from earlier pulses.

In the most recent large migration out of Africa, humans moved up to North Africa, some heading into West Africa, others crossing over to Eurasia via the Levant and into the Middle East, before advancing north-west to Europe. Meanwhile another group went over the Horn of Africa, before splitting up. Some people hugged the coast, heading to Asia, where some groups proceeded north from there, while others colonized South East Asia before moving down into Australia.

Other people crossed the Horn and took various routes north, some to Europe, others through China and north. Between 34,000 and 15,000 years ago, the northernmost reaches of the planet were drier and the now submerged Bering Land Bridge, linking Russia and North America, was passable. Humans traversed North America before crossing into South America, and advancing south. South America was the last continent colonized by Sapiens about 14,000 years ago.

It took many more years before Sapiens reached New Zealand, though, with the earliest evidence of human habitation there about 700 years ago (around 1300 AD).

11

JEWELS, BURIALS AND CEREMONY

In the late nineteenth century, it was much easier to decide what it meant to be human. We only knew about *Homo sapiens*, and didn't imagine that the remains of ancient ancestors were trapped in rock, waiting to be found. Those remains, and the artefacts associated with them, have torpedoed many of the assumptions we had about humans. Within palaeoanthropology, there is a schism about how to define a modern human: do we determine who is Sapiens based on their morphology and the physical capabilities written in their bones, or do we decide who belongs in our species based on their culture, making a value judgment about the traits that make us culturally human? This remains an open question.

We thought that only Sapiens were bipedal – an assumption that *Homo erectus* walked all over; we believed that art was the province of Sapiens – but then realized that Neanderthals may have made art and almost certainly created jewellery.

Similarly, our big brains marked us as Sapiens, but now we know that several of our *Homo* cousins had large brains too. For many years, language and sophisticated communication was the hallmark of humanity, but we now suspect that Neanderthals could also speak. Tool use was also considered a human trait, but the discovery of stone tools in eastern Africa has pushed the use of tools back 3.3 million years.

So what is it about Sapiens' behaviour and technology that makes it different from all the cultures that came before?

The oldest Acheulean tools – the distinctive pear-shared bifacial stone found in hand axes – date to about 1.7 million years ago. Between 320,000 and 305,000 years ago, populations in eastern Africa underwent a noticeable technological shift. People began to move away from the large, chunky Acheulean implements towards tools made of sharp flakes, which could be fashioned into blades and points. There's also a rise in the use of bone.

At Olorgesailie in Kenya, palaeoanthropologists have found obsidian tools whose raw black volcanic rock could have only come from miles away. They also discovered black and red rocks that had been used to create pigments, and the presence of more carefully crafted tools. These innovations speak to cognitively advanced capabilities, such as planning, enhanced technology, and possibly even a long-distance obsidian trade network (scholars are more sceptical of that claim, though).

This technology persisted until about 50,000 years ago, and in Africa is referred to as the Middle Stone Age. The transition to the Middle Stone Age also went hand in hand with changes in the landscape and mammal populations, and many larger species suddenly disappeared. The tool-makers would have hafted the stone or bone to a shaft, and artefacts in the Middle Stone Age show incremental, but consistent technological innovation. By contrast, Neanderthals' Mousterian technology remained fairly static and did not undergo such rapid transformation.

Projectile weapons were a distinctive feature of Sapiens' armoury. Surprisingly complex, they were made by securely attaching a sharpened point to a shaft. They offered a great advantage over previous thrusting spears – namely that the wielder was further away from a large, wounded animal or another aggressive human. Some refer to projectile weapons as an 'enabling technology', as it most likely allowed Sapiens a significant advantage over its prey and possibly over other human competitors.

About 500,000 years ago at Kathu Pan in South Africa, hominins made hafted, stone-tipped hunting spears, which they used to impale animals or rivals. The Schöningen Spears from Germany are about 300,000 years old. But throwing spears are a quantum leap in technological innovation, offering many benefits to the wielder. It is likely that such weapons meant that spear-throwing Sapiens did not suffer the impact traumas found on Neanderthal skeletons.

As early as 279,000 years ago, Sapiens in the Ethiopian Rift Valley were crafting pointed stone artefacts for use in projectile weapons. Scientists deduced this based on the artefacts' shape, microfractures and pattern of damage. This kind of activity becomes widespread later in the archaeological record, as more Sapiens benefited from the competitive advantage that such weaponry offered them.

For the most part, Sapiens' technological innovation resides in Africa. However, a bemusing discovery near Chennai in India has shown that individuals in other parts of the world also used sophisticated weapon-making techniques.[65] The cache of more than 7,000 stone artefacts such as blades and possibly points is at least 250,000 years old, much earlier than anyone thought such technology was feasible in the region. There are a few theories as to how this is possible: perhaps early humans developed such complex technology earlier than we thought, or maybe Sapiens left Africa sooner than we thought.

The Sapiens brain

About 7 million years ago, *Sahelanthropus tchadensis* had a brain slightly bigger than a can of Coca-Cola – about 370 cubic centimetres. And for 5 million years, hominins brains increased gradually, before we see a giant jump in brain size around 2 million years ago. We assume that this jump in brain volume, relative to body size, resulted in greater cognitive abilities, as at that time hominin technology becomes increasingly complex.

Modern humans have on average 86 billion brain neurons, which is more than double the number of other primates, but the source of our cognitive capacity extends beyond the number of neurons. A 2020 study of brain connections in 123 mammals, which included llamas, dolphins and, of course, humans, found that the patterns of links in the brain's connectome (the map of neural links) had roughly the same wiring design.[66] But an earlier study compared human and chimpanzee connectomes, and found that humans had thirty-three human-specific connections (about 6 per cent of the total).[67] These were longer and more important to the efficiency of the brain network than the other links that chimps and humans share.

Broca's area

In 1865, French physician Pierre Paul Broca described the cases of two patients who had lost their ability to speak properly and who both had lesions on the same part of the brain. The area became known as 'Broca's area'. It is in the frontal lobe of the human brain, and is implicated in language processing and communication. Scientists have shown that other parts of the brain are also involved in language processing, and they are now all grouped together in 'Broca's region'.

These longer connections create long-distance linkages across the brain, which may allow humans to integrate information from different parts of the brain more efficiently.

A noticeable difference between chimps and humans is the more specialized connectivity in an area of the human brain called Broca's area. The authors of the study suspect that the wiring density in human-language processing areas contributed to the evolution of complex language in the human lineage.

But Sapiens' and Neanderthals' brains were approximately the same size – so what made Sapiens different and allowed us to achieve things that our cousins couldn't?

One reason may be the shape. Our brains are much rounder, while Neanderthals – and others before them – had rather elongated head shapes. The argument is that our skull changed shape to accommodate certain regions of the brain as they evolved. Sapiens have a larger cerebellum, which means 'little brain'. This area, which is located at the base of your skull, governs voluntary movement, balance, motor skills and speech. And a 2022 study suggests that a mutation in the TKTL1 gene allowed our Sapiens ancestors to produce extra neurons in their frontal lobe.[68] The frontal lobe, which sits behind our forehead, is the largest of our brain's four lobes. It is involved in higher-level executive functions, such as planning, impulse control and social interaction, among many others. However, in a 2023 response, other scientists disputed the study's finding, saying that the putative Neanderthal variant can be found in modern human populations.

So while Sapiens and Neanderthals may have had similar brain sizes, their shape and organization were different – and the differences in those of Sapiens gave them a competitive advantage over their cousins.

The development of complex culture

While scientists argue about the cognitive faculties of Neanderthals and how to attribute cultural artefacts to them, we do not have that problem with Sapiens. We know that our species was and is capable of symbolic complex thought.

We're not sure precisely where it started, though. It also appears to be a trait that developed with time, rather than sprouting fully formed. The earliest Sapiens that we've found so far were in Morocco, Ethiopia and South Africa, but we have not uncovered particularly complex technology alongside these remains. This absence highlights that physical evolution did not necessarily go hand in hand with cognitive development.

While the oldest Sapiens remains that we've discovered are

about 315,000 years old, it is another 100,000 years before we start seeing behavioural modernity in the archaeological record. At the moment, many of those finds seem to be centred in South Africa, either at rich sites inland in the KwaZulu-Natal province or along its south-eastern coast.

In Border Cave in KwaZulu-Natal, ancient Sapiens intentionally laid down matting for beds 200,000 years ago. The grasses were interspersed with charcoal from the broad-leafed camphor bush, which would have repelled crawling insects and stopped them from burrowing into the bedding. Meanwhile, 120,000 years ago in Klasies River, which is about 1,000 kilometres south of KwaZulu-Natal along South Africa's east coast, humans lived in the caves that dot the coastline. And researchers have found the oldest evidence that people were roasting tubers. The carbohydrates within the starches are considered vital for brain development.

Further inland in the Kalahari Desert, ancient Sapiens collected twenty-two white calcite crystals and brought them to the Ga-Mohana Hill rock shelter about 105,000 years ago. The crystals don't appear to have had a practical use but were intentionally taken to the shelter, and were found alongside stone tools, shells, bones with the telltale signs of butchery, and ostrich-egg fragments. The discoverers believe that people used the shells as water vessels. Back then, the Kalahari would have been green and lush.

Around the same time in Blombos Cave on South Africa's south coast, long-ago people created an art processing workshop, and produced a liquid ochre mixture which they stored in abalone shells. Scientists have dated the artefacts to 100,000 years old. Alongside this paint kit, researchers also found the tools needed to produce it: ochre, bone, charcoal, grindstones and hammer stones.

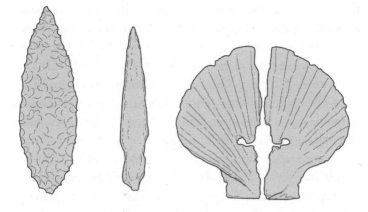

The southern Cape region of South Africa is home to some of the earliest evidence of complex culture in *Homo sapiens*.

From about 77,000 years ago, there is a spurt of innovation and complexity in the South African archaeological record. The innovations include bone tools, beads, ochre, ostrich shells, and the manufacture of glue using more than one ingredient. Two notable techno-complexes were based on initial finds at Still Bay and Howiesons Poort, although examples of these technologies have since been discovered at many other locations. The former culture, observed at sites in coastal areas in South Africa, is characterized by bifacial points. The latter involved the almost exclusive use of blades and bladelets, although peculiarly many artefacts from the 'post-Howiesons Poort' period lack the innovation seen in this industry.[69]

About 75,000 years ago, the inhabitants of Blombos Cave strung shells as beads and scored slabs of engraved ochre with criss-cross geometric patterns. At Pinnacle Point, also on the south coast, humans were systematically heat-treating stone materials to enhance their flaking properties to improve their tool-making.

Meanwhile, the oldest known possible stone and arrow heads come from the Sibudu Cave in KwaZulu-Natal (dated to 64,000 and 61,000 years old respectively), as well as the earliest needle (61,000 years old). Sibudu contains instances of the Howiesons Poort technology.

There are also examples of supposedly 'modern' behaviour in other places in Africa and the Middle East, but none of them has been quite as fecund as the sites in southern Africa.

In North Africa, at sites spanning 130,000 to 60,000 years ago, archaeologists have found stone tools not present in South Africa, such as stone implements that had stems to which their makers attached handles. Such technology is not found further south and, conversely, the heat-treated thin blades found in the south were not present in the north.

In Skhul, Israel, archaeologists discovered perforated shells along with the remains of early Sapiens in a 100,000- to 135,000-year-old sedimentary layer. The shells are less than 2 centimetres long, and do not belong in the ancient cave according to archaeologists. The shells all had similar holes and had been originally occupied by marine creatures, even though Skhul is relatively far from the sea. Consequently, the researchers argue that they were deliberately selected, perforated and brought to Skhul. Similarly, a site in Oued Djebbana, Algeria was never closer than 190 kilometres to the sea during that geological epoch (the Upper Pleistocene). The shells found there are thought to be about 100,000 years old.

In Morocco's Grotte des Pigeons, the evidence is less debatable. At that site, people were stringing shells together, most likely for jewellery, 82,000 years ago. Similar to the shell beads discovered at Blombos (which are about 7,000 years younger), they were covered in red ochre. Many argue that the existence of beads at several geographically distant sites suggests that there was an enduring and widespread bead-working tradition in Africa and the Levant.

Identifying cultural modernity in other places becomes more tricky, though. In Europe, we cannot ascribe specific artefacts or even artwork to a specific group. Additionally, we cannot know if existing *Homo* groups honed these technologies and altered their behaviours based on their interactions with Sapiens. But, as some scholars argue, if other humans were able to learn these behaviours, it means that they are not Sapiens-specific and that other species were able to act in a 'modern' way.

From about 50,000 years ago, we begin to see the remnants of Sapiens' cultural activities in the form of art, tools and jewellery, but they could have been there much earlier, as the Apidima 1 remains in Greece suggest.

Behavioural modernity in Asia is even more complicated, because there may have been a number of human groups in Asia around the same time: Denisovans, *H. floresiensis* and *H. luzonensis*, as well as the newly discovered *H. longi* in China (although many dispute that the 'Dragon Man' is indeed a new species).

We do know, however, that Sapiens were present in Asia by about 40,000 years ago, and at an archaeological site in Xiamabei in northern China dated to that time, someone created a small ochre-processing workshop, replete with bladelet-like tools that bore traces of hafting, and a bone tool. So far, they are the earliest known examples of pigment-making and such tools in Asia.

Intentional burial

About 78,000 years ago, someone dug a shallow circular grave at the Panga ya Saidi cave mouth in Kenya. In it, they placed a two-year-old child on their right side, with their knees drawn toward their chest. Scientists found the toddler, named Mtoto, which means 'child' in the local Swahili language, 3 metres below the current cave floor.

Their discoverers say that the position of their ribs and spine suggest that they decomposed within the pit. Their tightly bound foetal position, the scientists assert, was held in place by a shroud, and their head was once supported by a makeshift pillow.[70] In other words, they argue that someone intentionally buried Mtoto.

Mtoto was discovered in a sedimentary layer that contained Middle Stone Age tools, which were synonymous with *Homo sapiens*. Additionally, their teeth mark them as Sapiens.

This is the oldest example of intentional burial in Africa that we know of, and is a distinctly human behaviour. (The jury is still out on whether Neanderthals intentionally buried their dead, and the matter is far from decided.) Palaeoanthropologists have long been bemused at the relative dearth of funeral sites in Africa, where our genus originated.

From about 800,000 years ago, the archaeological record contains a variety of ways in which hominins handled their dead. For some early hominins, this may have involved cannibalism. At Gran Dolina in Spain, we suspect that about 780,000 years ago *H. antecessor* individuals cannibalized several juveniles and children. In the same Atapuerca mountain range, early Neanderthals appear to have disposed of bodies in a deep cave, Sima de los Huesos, about 300,000 years later. Thousands of kilometres away in South Africa's Rising

Star Cave, numerous *H. naledi* individuals were deposited in a narrow cave between 335,000 and 236,000 years ago. The scientific team that discovered the specimens believe that *H. naledi* purposefully buried its dead, although these findings still need to be peer-reviewed.

Sima de los Huesos
500-400,000 years ago

Shanidar Cave
70-60,000 years ago

El Sidrón
approx 49,000 years ago

Gran Dolina
850-780,000 years ago

Tabun C1
approx 122,000 years ago

Taramsa Hill
approx 69,000 years ago

Skhul and Qafzeh
130-90,000 years ago

Herto
160-150,000 years ago

Bodo
approx 600,000 years ago

Panga ya Saidi
approx 78,000 years ago

● *Homo sapiens*
■ Neanderthals
◆ Other *Homo* hominins

Rising Star Cave
335-236,000 years ago

Border Cave
approx 74,000 years ago

While body disposal has been recorded at numerous sites around the world, intentional funeral burial is thought, at the moment, to be something that only Sapiens are capable of.

However, funeral practices are something else entirely, and are a particularly fraught topic in a discipline characterized by disagreement. Intentional burial goes beyond hiding bodies from scavengers or avoiding the presence of a rotting body nearby. By deliberately burying a body, an individual or group invests additional time and resources for reasons other than necessity. Such actions can include placing the body in a specific position, wrapping them or including a keepsake. This type of conduct elevates the practice above simple body disposal and into the realm of symbolism and behaviour that is currently considered the sole ambit of Sapiens.

Prior to the discovery of Mtoto, a six-month-old child found in Border Cave in South Africa was the oldest known human burial in Africa. The infant was entombed in 74,000-year-old sedimentary layers, making it a few millennia younger than Mtoto. What sets the Border Cave infant apart from previous suspected instances of intentional burial was that someone interred them with the shell of a sea snail, *Conus bairstowi*. The shell had distinctive marks, indicating that it was once regularly worn around someone's neck.

However, Mtoto and the Border Cave child are not the oldest burials in the human record – they were the oldest intentional burials in Africa. A mass grave site in Qafzeh, Israel is the first example of intentional burial. The remains of fifteen anatomically modern humans, dated to about 92,000 years, were found in the Qafzeh Cave. More than half of them were children. One pre-teen boy was placed in a rectangular grave, with his elbows hugging his body and his hands clasping his neck. He was buried with deer horns next to his chest. At the Qafzeh Cave, researchers also found more than seventy pieces of ochre, which are thought to have been used in funeral practices.

From what we have found so far, there is a disproportionate representation of child graves in the Sapiens archaeological record. We are not sure why, but some scholars have flagged that many hunter-gatherer communities view child death as 'unnatural' and so there may have been special efforts made in the funeral rites of children.

As far as we know, funeral practices remain a distinctly Sapiens behaviour. The intentional and symbolic act of saying farewell to the dead continues to be a hallmark of being a modern human, and distinguishes us from all the ancestors in our lineage.

12 THE BIG QUESTIONS

Between 300,000 and 100,000 years ago, many human species walked the planet. But today, there is only one left – *H. sapiens* – and we don't know why. This is one of the big questions in human evolution: What happened to everyone else?

H. naledi walked the wooded grasslands of the Cradle of Humankind in South Africa. Early *H. sapiens* were present in modern-day Morocco, while *H. rhodesiensis* (also known as *H. bodoensis* or *H. heidelbergensis*) was alive in central and southern Africa. Heavy-browed Neanderthals were colonizing Europe and interacting with Denisovans in Asia. *H. luzonensis* and the diminutive *H. floresiensis* eked out an existence on their islands, while *H. erectus* still wandered parts of Indonesia and *H. longi* lived in China. There may have even been more species we didn't know about – especially if we consider the ghost DNA in our genetic code. But by about 40,000 years ago, Sapiens is the last human standing out of what was a large and diverse group of bipedal hominins.

The major upheaval in the existence of these extinct humans may have been the arrival of Sapiens, whose presence – for several possible reasons – may have led to their demise. One hypothesis is that aggressive Sapiens with their advanced technologies killed local human populations when they encountered them. While that did probably happen in some instances, there is mounting evidence that this was not the case in general.

For one thing, Sapiens appears to have arisen from multiple populations within Africa, and would have encountered other Sapiens and interbred with them even before they left the continent. These Sapiens populations lived thousands of kilometres apart and had adapted to their distinct environments, both biologically and technologically. When these groups came into contact, they would have been able to share their innovations and culture, as well as genetic material.

However, it is possible that limited resources as a result of changing climatic conditions could have put Sapiens in direct competition with other species. Or Sapiens may simply have been more resilient in the face of these environmental fluctuations, while others may have perished.

The 'climate change' hypothesis of human extinction is becoming increasingly popular. One group of researchers believe that 'laziness' coupled with an inability to adapt to new conditions were what forced the demise of *Homo erectus* in Asia. According to them, *H. erectus* relied on convenient stones for tools, which did not really evolve during this long-enduring species' history. As resources became scarcer, the species was unable to adapt and so died out.

A 2022 study modelled ancient climates and the niches in which *H. erectus*, *H. heidelbergensis* and Neanderthals lived, and suggested that these species lost a significant portion of their environmental comfort zone before they went extinct.[71] For all species, the climate became quite a lot colder, and turned drier for *H. heidelbergensis* and Neanderthals, and wetter for *H. erectus*. This would have had a knock-on effect on their food sources, which may have dwindled or, in the case of animals, migrated.

H. erectus and *H. heidelbergensis* lost about half of their niche to climate change, while that of Neanderthals was reduced by about a quarter. With increased competition for resources, craftier and more resilient Sapiens – with our innovative technologies, such as throwing spears and arrows – may have had the upper hand, either because we were better equipped to kill human rivals or because we could outcompete them.

Some scholars also think that Sapiens' longer, lighter legs gave us an advantage over other humans. Neanderthals, with their shorter leg bones and longer Achilles' tendons, are morphologically more suited to sprinting and hiking up hills, and so they were better adapted to ambushing their prey in the cold forested areas where they lived. Sapiens, on the other hand, have a loping stride that can cover long distances. If a changing climate pushed Neanderthals out of hilly areas and into an open landscape, they would have had to adapt their hunting strategies, and loping Sapiens may have ultimately been the more successful of the two groups.

Another possibility is that Sapiens interbred other species out of existence. There is now a great deal of evidence that Sapiens had offspring with Neanderthals and Denisovans, as there are traces of their genetics in our DNA. Some people in Eurasia have up to 2 per cent of Neanderthal DNA, while Melanesians have up to 5 per cent Denisovan DNA. A mystery human ancestor contributed between 2 and 19 per cent of their genetic ancestry to people living in West Africa. A puzzling corollary to this is that out of the more than thirty Neanderthal genomes we've sequenced so far, none of them contains Sapiens DNA. That could simply be a sampling bias. At the same time, and rather strangely, modern humans do not contain Neanderthal mitochondrial DNA. One theory is that, since mtDNA is inherited along the maternal line, only male Neanderthals and female Sapiens could have live offspring, and that male hybrids may have been less fertile.

In the case of Neanderthals, it is possible that congenital diseases made them sickly, reduced their numbers and ultimately made them less competitive than Sapiens. There is also genetic evidence that Neanderthals lived in small communities, with little migration between populations. For example, the parents of the Altai Neanderthal girl from the Denisovan Cave were likely half-siblings, and that this was not a one-off event in her lineage. If such inbreeding were common among Neanderthals, it would have made them vulnerable to genetic diseases.

Nevertheless, continuous absorption of archaic human DNA into Sapiens could have resulted in the disappearance of other species.

Like in Africa many years before, such assimilation – both biological and cultural – could also have given Sapiens a competitive advantage. Local human species had been living in their regions for thousands of years, and presumably had adapted to local conditions. Interbreeding Sapiens could have picked up evolved traits and cherry-picked behaviours that made them better able to survive in a variety of climates and environments. Such resilience would have made Sapiens better adapted to changing climates too.

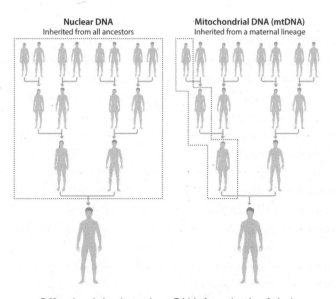

Nuclear DNA
Inherited from all ancestors

Mitochondrial DNA (mtDNA)
Inherited from a maternal lineage

Offspring inherit nuclear DNA from both of their parents, but mitochondrial DNA is only passed from mothers to their children.

Are humans still evolving?

In short, yes. Every time a child is born, they contain small mutations in their DNA. Such mutations are the building blocks of species evolution. The questions are, rather, how are we changing and how quickly this is happening?

Several forces drive evolution, including mutation, natural selection, genetic drift and gene flow. In natural selection, individuals who were better adapted to their environment were more likely to survive and have children. Genetic drift occurs in a population when the frequency of an existing gene variant changes due to random chance (making a once-common variant rarer or an uncommon one more prevalent), and gene flow occurs when different populations interbreed. However, modern humans have other evolutionary tools – technology and culture. By being able to manipulate our environment and bodies, we are changing how we evolve.

Scientists agree that we are evolving more rapidly than ever before in our history. A 2007 study found that the pace of human evolution has accelerated in the last 40,000 years, particularly since we started farming 10,000 years ago.[72] Part of the reason is that there are quite simply more of us. There are now more than 8 billion humans on the planet, many of whom are having children who are born with a handful of mutations. Most mutations are just a blip in the code of life; very few of them will actually benefit the person. However, interesting things happen when populations find themselves in new environments. In such situations, larger populations are able to respond more rapidly in the form of new mutations. But even though the scale of human migration is greater than it has ever been in our history, we are not merging into a single homogeneous gene pool, and populations in different places are becoming increasingly genetically different.

Also, technological and cultural innovations – such as antibiotics and clean water – mean that we are living longer, and so are having children later in life. The children of older men tend to have more genetic mutations. Some scholars believe that as Sapiens' lifespans continue to lengthen, we may evolve to extend child-bearing age.

In our recent history, numerous populations have uniquely adapted to their environments, evolving in ways that other groups haven't. Indigenous Siberian populations, for example, are better equipped to survive very cold temperatures: they are able to generate heat in their brown fat tissue and, in certain circumstances, their bodies open blood vessels, known as vasodilation, to increase the flow of warm blood near the skin and stave off frostbite.

Meanwhile, in South East Asia, the Bajau people can dive more than 70 metres underwater to collect shellfish or spear fish in a single breath. They can hold their breath for many minutes at a time, while the average person can manage about 30 to 90 seconds. A change to the Bajau's PDE10A gene means that their spleens are almost double the size of other groups living in the region, creating a reservoir for them to store oxygen-rich red blood cells.

Another adaptation is the ability to drink milk. When we are born, almost all humans are able to produce lactase, an enzyme which breaks down lactose in milk. We are the only animal that retains this ability in adulthood – although not everyone can do it. When ancient populations in sub-Saharan Africa, Arabia and Europe began keeping cattle and camels for milk, their genes – independently – mutated to allow them to keep producing lactase into adulthood. About two-thirds of people have some form of lactose intolerance.

Similarly, scientists believe that pale skin is a relatively recent adaptation. When human ancestors lost their hair to better regulate their temperature, they also developed darker skin. Dark skin offers protection against the UV radiation and DNA damage, which is something that people who live in particularly sunny places, such as Africa and around the equator, require. But as Sapiens populations moved north into colder climates, they had less need for UV protection, but a greater need to produce their vitamin D. The body creates vitamin D from sunlight. Lighter skin is better at making vitamin D, which is necessary to regulate the amount of calcium and phosphate in the body.

The presence of other human DNA may also play a role.

A 2017 study found that Neanderthal DNA affects skin tone, hair colour, height and a number of other metrics in present-day Europeans.[73] However, it is not a simple correlation: they found multiple mutations at different places contributed to skin and hair colour in Europeans, and that genetic variations resulted in both lighter and darker skin and hair colours. This points to variability within Neanderthals themselves.

Disease prevalence also forces evolutionary pressures on populations, and there are numerous examples of people developing genetic resistance to diseases. For example, populations that moved into urban areas long ago are more likely to have a natural resistance to tuberculosis and leprosy than recently urbanized populations. In places where malaria is or was endemic, some people have sickle-shaped red blood cells which are less hospitable homes for the malaria-causing parasite. Scientists have linked this mutation to a girl who lived in West Africa about 7,300 years ago.

Children who had these mutations were able to grow up, reproduce and continue their lineage – passing on the mutation that allowed them to survive. However, the mutation came at a terrible cost. When people who both have this mutation have children, their kids have a one-in-four chance of contracting sickle-cell disease, in which their red blood cells have unusual shapes. In the most severe cases involving sickle-cell anaemia, their red blood cells break apart leaving their bodies starved of oxygen.

In the past, survival favoured those, such as the West African girl with sickle-shaped blood cells, while community members who did not have such advantages died. Today, the opposite is also true: harmful mutations do not necessarily kill people, and so these mutations remain in the gene pool. For example, a heart defect caused by a genetic disorder can be fixed with surgery, or the person may even be able to have a heart transplant. Even more ubiquitous, glasses give people with poor eyesight independence and autonomy, something that would not have been possible thousands or even a few hundred years ago. There is less selective pressure on having 20/20 vision.

Epigenetics

In some instances, a shifting environment can have a rather speedy impact on our genetics. The genes themselves are not altered, but rather which genes are expressed and when. This field, known as epigenetics, is a burgeoning area of research. Unlike genetic changes, epigenetic changes can be reversed.

Epigenetic changes can also be passed on to offspring. A 2014 study found that newborns whose mothers reported smoking in their first trimester had epigenetic modifications to their DNA, which were not present in babies born to non-smokers.[74]

However, scientists are not sure how epigenetic mutations persist within populations and over generations and inform genetic variation.

Vaccines accelerate the process of immunity acquisition. So instead of waiting for genetic resistance traits to be bred into a population over the course of decades, centuries or millennia, vaccines can immediately stop people, particularly young children, from dying from many diseases, such as measles and diphtheria. The World Health Organization says that childhood vaccinations are one of our most successful and cost-effective public health interventions, and have saved millions of children's lives.

It is also possible that the rising prevalence of Caesarean sections is changing both the size of babies' heads and women's pelvises. Before this life-saving medical intervention, women with narrow hips and babies with large heads struggled to survive childbirth. Some research suggests that C-sections are allowing such genes – those for big newborn heads and narrow pelvises – to remain in the gene pool.[75] Not all scientists are convinced, though, and it may be many years before we can be certain that this is in fact a trend.

The practice of gene-editing could also reduce the prevalence of certain diseases within human populations, but it is very controversial and there are questions about its safety and morality.

Culture offers evolutionary advantages

Humans also adapt alongside their culture, and our current rapid technological advances, from artificial intelligence to personalized medicine, could accelerate our evolution.

But innovations that we now take for granted – such as improved sanitation and clean water – promote human evolution. Because of such interventions, we are more likely to reach adulthood and have children. This increases the number of humans within the population, and with it the prevalence of genetic mutations.

Our culture and lifestyles are also changing our bodies. Sapiens evolved to be light-boned and fleet-footed, which may have given us a competitive advantage over other *Homo* species. But our modern sedentary lifestyles – something which became more common with the advent of farming and continues with our time behind desks and steering wheels – is decreasing our bone density, resulting in a higher likelihood of bone fractures, particularly in the hip.

Our brains also appear to be getting smaller. Neanderthals and Sapiens at one point had similar-sized brains, but our brains appear to have reached peak size between 20,000 and 10,000 years ago, just before the advent of farming. We're not sure why. Some scholars suggest that it is because we began to eat less fat, getting our energy from more starches. Fat is strongly implicated in brain development, so when we started eating less of it, our brains got smaller. Others say that our reliance on farming meant that droughts and other natural disasters have led to bouts of famine. In such a situation, expending energy on larger brains may have been wasteful, and so we evolved smaller brains. However, a 2021 paper argued that in the last few thousand years, we began to increasingly rely on external knowledge and group-level decision-making. As a result, we didn't need such big brains, with their associated energy cost.

AFTERWORD

Humans are remarkable creatures. Some of our simplest actions – such as walking upright, holding a pen to write down a phone number, and riding a bicycle – are only possible because of how our lineage has adapted and survived over the course of millennia. We take this for granted, because it seems so quotidian, so banal.

Our mental faculties developed alongside our physical adaptations, allowing us the introspection and wonder to ask questions, such as: who are we? Why are we here? Our amazing brains allow us to record our histories, unearth forgotten stories, and assimilate all of this information into our understanding of the world and our place in it.

The investigation into our origins is perhaps one of our most human activities. We use many of the skills and adaptations that make us singularly human – curiosity, innovation, technology, social engagement, imagination and empathy. And by tracking our inquiries over the centuries, we can also see how much we as a species have changed.

The search to understand who we are and where we come from is a uniquely human enterprise and, like us, our story continues to evolve and change.

TIMELINE OF DISCOVERIES

1848   *Homo neanderthalensis* (Gibraltar skull) –
Rock of Gibraltar

1856   *Homo neanderthalensis* (Neanderthal 1 skullcap) –
Neander Valley, Germany

1868   *Homo sapiens* (early modern humans) –
Cro-Magnon, France

1891   *Homo erectus* (Java Man) – Trinil, Java

1908   *Homo heidelbergensis* – Mauer, Germany

1912   Piltdown Man hoax

1921   *Homo rhodesiensis* (Broken Hill skull) –
Kabwe, Zimbabwe

1924   *Australopithecus africanus* (Taung child) –
Taung, South Africa

1929   *Homo erectus* (Peking Man) – Zhoukoudian, China

1938   *Paranthropus robustus* – Kromdraai, South Africa

1959   *Paranthropus boisei* – Olduvai Gorge, Tanzania

1964   *Homo habilis* – Olduvai Gorge, Tanzania

1972   *Homo rudolfensis* – East Turkana, Kenya

1974   *Australopithecus afarensis* (Lucy) – Laetoli, Tanzania

1975   *Homo ergaster* – Koobi Fora, Kenya

1978   *Homo daliensis* (Dali Man) – Shaanxi, China

1995   *Australopithecus anamensis* – West Lake Turkana,
Kenya

1995   *Ardipithecus ramidus* – Afar region, Ethiopia

1996   *Australopithecus bahrelghazali* – Bahr el Gazel, Chad

1997 *Homo antecessor* – Gran Dolina, Spain
1999 *Australopithecus garhi* – Bouri, Ethiopia
2001 *Kenyanthropus platyops* – Lomekwi, Kenya
2001 *Orrorin tugenensis* – Tugen Hills, Kenya
2001 *Ardipithecus kadabba* – Middle Awash, Ethiopia
2002 *Sahelanthropus tchadensis* – Toros-Menalla, Chad
2004 *Homo floresiensis* – Flores, Indonesia
2010 *Australopithecus sediba* – Malapa, South Africa
2010 Denisovans – Denisova Cave, Siberia
2015 *Homo naledi* – Rising Star cave, South Africa
2015 *Australopithecus deyiremeda* – Afar region, Ethiopia
2019 *Homo luzonensis* – Luzon, the Philippines
2021 *Homo longi* – Harbin, China

NOTE: This list reflects the complexity of palaeoanthropology. Sometimes specimens are discovered years before they are described. In other instances, they are shifted from one species designation to another. There is also a lag between the discovery of a specimen and it being officially described. Nevertheless, this list is designed to give you a general idea of when different hominins entered the discussion and where they came from.

TIMELINE OF HUMAN ANCESTORS

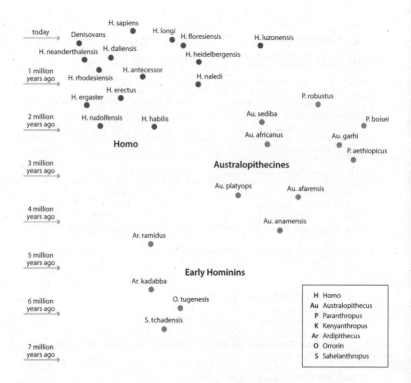

today

H. sapiens
Denisovans
H. longi
H. floresiensis
H. luzonensis
H. neanderthalensis
H. daliensis
H. heidelbergensis

1 million
years ago
H. rhodesiensis
H. antecessor
H. naledi
H. erectus
H. ergaster

2 million
years ago
H. rudolfensis
H. habilis
P. robustus

Au. sediba
P. boisei
Au. africanus
Au. garhi
P. aethiopicus

Homo

Australopithecines

3 million
years ago

Au. platyops
Au. afarensis

4 million
years ago

Au. anamensis

Ar. ramidus

5 million
years ago

Early Hominins

Ar. kadabba

6 million
years ago
O. tugenesis

S. tchadensis

7 million
years ago

H	Homo
Au	Australopithecus
P	Paranthropus
K	Kenyanthropus
Ar	Ardipithecus
O	Orrorin
S	Sahelanthropus

ENDNOTES

1 PRIMATES

1 'A parapithecid'stem anthropoid of African origin in the Paleogene of South America' by Erik R. Seiffert, Marcelo F. Tejedor *et al.* (*Science*, Vol. 368, No. 6487, pp. 194–97, April 2020).

2 A SHREWDNESS OF APES

2 'The genetic basis of tail-loss evolution in humans and apes' by Bo Xia, Weimin Zhang *et al.* (*bioRxiv*, https://doi.org/10.1101/2021.09.14.460388, September 2021).

3 'Basal Primatomorpha colonized Ellesmere Island (Arctic Canada) during the hyperthermal conditions of the early Eocene climatic optimum' by Kristen Miller, Kristen Tietjen and K. Christopher Beard (*PLoS ONE*, https://doi.org/10.1371/journal.pone.0280114, January 2023).

4 'Initial sequence of the chimpanzee genome and comparison with the human genome' by Robert H. Waterson, Eric S. Lander and Richard K. Wilson (*Nature*, Vol. 437, pp. 69–87, September 2005).

5 'Zoonomia' by Sacha Vignieri (*Science*, Vol. 380, Issue 6643, pp. 356–57, April 2023).

6 'Great apes use self-experience to anticipate an agent's action in a false-belief test' by Fumihiro Kano, Christopher Krupenye, Satoshi Hirata, Masaki Tomonaga, and Josep Call (*PNAS*, Vol. 116, No. 42, September 2019).

7 'Biochronology of South African hominin-bearing sites: a reassessment using cercopithecid primates' by Stephen R. Frost, Frances J. White, Hailay G. Reda and Christopher C. Gilbert (*PNAS*, Vol. 119, No. 45, October 2022).

3 THE LEAP FROM APE TO HOMININ

8 'A new hominid from the Upper Miocene of Chad, Central Africa' by Michel Brunet, Franck Guy *et al.* (*Nature*, Vol. 418, pp. 145–51, July 2002).

9 'Nature and relationships of Sahelanthropus tchadensis' by Roberto Macchiarelli, Aude Bergeret-Medina, Damiano Marchi and Bernard Wood (*Journal of Human Evolution*, Vol. 149, December 2020).

10 'Postcranial evidence of late Miocene hominin bipedalism in Chad' by Franck Guy, Guillaume Daver *et al.* (*HAL*, https://hal.science/hal-03037386, January 2021).

11 *Human Evolution: A Very Short Introduction* by Bernard Wood (Oxford University Press, 2019).

4 THE SOUTHERN APE

12 'U–Pb-dated flowstones restrict South African early hominin record to dry climate phases' by Robyn Pickering, Andy Herries *et al.* (*Nature*, Vol. 565, pp. 226-229, January 2019).

13 'New genetic and morphological evidence suggests a single hoaxer created "Piltdown man"' by Isabelle De Groote, Linus Girdland Flink *et al.* (*Royal Society Open Science*, Vol. 3, Issue 8, August 2016).

14   'Australopithecus africanus: the man-ape of South Africa' by Raymond A. Dart (Nature, Vol. 115, pp. 195–99, February 1925).

15   'Eagle involvement in accumulation of the Taung child fauna' by L. R. Berger, R. J. Clarke et al. (Journal of Human Evolution, Vol. 29, Issue 3, pp. 275–99, September 1995).

16   'Strontium isotope evidence for landscape use by early hominins' by Sandi R. Copeland, Matt Sponheimer et al. (Nature, Vol. 474, pp. 76–78, June 2011).

6 TOOLS FOR TINKERING

17   'Compound tool construction by New Caledonian crows' by A. M. P. von Bayern, S. Danel et al. (Scientific Reports, Vol. 8, October 2018).

18   'Wild monkeys flake stone tools' by Tomos Proffitt, Lydia V. Luncz et al. (Nature, Vol. 539, pp. 85–88, October 2016).

19   'Earliest Olduvai hominins exploited unstable environments ~ 2 million years ago' by Julio Mercader, Pam Akuku et al. (Nature Communications, Vol. 12, January 2021).

20   'Expanded geographic distribution and dietary strategies of the earliest Oldowan hominins and Paranthropus' by Thomas W. Plummer, James S. Oliver et al. (Science, Vol. 379, Issue 6632, pp. 561–66, February 2023).

7 WHAT IT MEANS TO BE HUMAN

21   'Bipedal steps in the development of rhythmic behaviour in Humans' by Matz Larsson, Joachim Richter et al. (Music & Science, Vol. 2, December 2019).

22   'Divergence-time estimates for hominins provide insight into encephalization and body mass trends in human evolution' by Hans P. Püschel, Ornella C. Bertrand et al. (Nature Ecology & Evolution, Vol. 5, pp 808–19, April 2021).

23   'Early Homo at 2.8 Ma from Ledi-Geraru, Afar, Ethiopia' by Brian Villmoare, William H. Kimbel et al. (Science, Vol. 347, Issue 6228, pp. 1352–55, March 2015).

24   'Chimpanzee genomic diversity reveals ancient admixture with bonobos' by Marc de Manuel, Martin Kuhlwilm et al. (Science, Vol. 354, Issue 6311, pp. 477–81, October 2016).

25   'Late Pliocene environmental change during the transition from Australopithecus to Homo' by Joshua R. Robinson, John Rowan et al. (Nature Ecology & Evolution, Vol. 1, May 2017).

26   'Impact of meat and Lower Palaeolithic food processing techniques on chewing in humans' by Katherine D. Zink and Daniel E. Lieberman (Nature, Vol. 531, pp. 500–503, March 2016).

27   'Homo erectus at Trinil on Java used shells for tool production and engraving' by Josephine C. A. Joordens, Francesco d'Errico et al. (Nature, Vol. 518, pp. 228–31, February 2015).

8 A SPECIES TO TAKE OVER THE WORLD

28   'Contemporaneity of Australopithecus, Paranthropus, and early Homo erectus in South Africa' by Andy I. R. Herries, Jesse M. Martin et al. (Science, Vol. 368, Issue 6486, April 2020).

29   'New hominin remains and revised context from the earliest Homo erectus locality in East Turkana, Kenya' by Ashley S. Hammond, Silindokuhle S. Mavuso et al. (Nature Communications, Vol. 12, April 2021).

30   'The earliest Pleistocene record of a large-bodied hominin from the Levant supports two out-of-Africa dispersal events' by Alon Barash, Miriam Belmaker, et al. (Scientific Reports, Vol. 12, February 2022).

31   'Hominin occupation of the Chinese Loess Plateau since about 2.1 million years ago' by Zhaoyu Zhu, Robin Dennell et al. (Nature, Vol. 559, pp. 608–12, July 2018).

32   'Dating the skull from Broken Hill, Zambia, and its position in human evolution' by Rainer Grün, Alistair Pike et al. (Nature, Vol. 580, pp. 372–75, April 2020).

33   'Evidence for the cooking of fish 780,000 years ago at Gesher Benot Ya'aqov, Israel' by Irit Zohar, Nira Alperson-Afil et al. (Nature Ecology & Evolution, Vol. 6, pp. 2016–28, November 2022).

34   'The discovery of fire by humans: a long and convoluted process' by J. A. J. Gowlett
      (*Philosophical Transactions of the Royal Society B*, Vol. 371, Issue 1696, June 2016).

35   'Fire as an engineering tool of early modern humans' by Kyle S. Brown, Curtis W. Marean *et al.*
      (*Science*, Vol. 325, Issue 5942, pp. 859–862, August 2009).

9 OUR CLOSEST COUSINS

36   'Morphology, pathology, and the vertebral posture of the La Chapelle-aux-Saints Neandertal' by
      Martin Haeusler, Erik Trinkaus *et al.* (*PNAS*, Vol. 116, No. 11, pp. 4923–27, February 2019).

37   'Articulatory capacity of Neanderthals, a very recent and human-like fossil hominin' by Anna
      Barney, Sandra Martelli *et al.* (*Philosophical Transactions of the Royal Society B*, Vol. 367, Issue
      1585, pp. 88–102, January 2012).

38   'Neanderthals and Homo sapiens had similar auditory and speech capacities' by Mercedes
      Conde-Valverde, Ignacio Martínez *et al.* (*Nature Ecology & Evolution*, Vol. 5, pp. 609–15,
      March 2021).

39   'Differential DNA methylation of vocal and facial anatomy genes in modern humans' by David
      Gokhman, Malka Nissim-Rafinia *et al.* (*Nature Communications*, Vol. 11, March 2020).

40   'A mitochondrial genome sequence of a hominin from Sima de los Huesos' by Matthias Meyer,
      Qiaomei Fu *et al.* (*Nature*, Vol. 505, pp. 403–406, January 2014).

41   'Nuclear DNA sequences from the Middle Pleistocene Sima de los Huesos hominins' by
      Matthias Meyer, Juan-Luis Arsuaga *et al.* (*Nature*, Vol. 531, pp. 504–507, March 2016).

42   'Dental evolutionary rates and its implications for the Neanderthal–modern human divergence'
      by Aida Gómez-Robles (*Science Advances*, Vol. 5, Issue 5, May 2019).

43   'Using genetic evidence to evaluate four palaeoanthropological hypotheses for the timing of
      Neanderthal and modern human origins' by Phillip Endicott, Simon Y. W. Ho and Chris
      Stringer (*Journal of Human Evolution*, Vol. 59, Issue 1, pp. 87–95, July 2010).

44   'Modeling Neanderthal clothing using ethnographic analogues' by Nathan Wales (*Journal of
      Human Evolution*, Vol. 63, Issue 6, pp. 781–95, December 2012).

45   'Direct evidence of Neanderthal fibre technology and its cognitive and behavioral implications'
      by B. L. Hardy, M.-H. Moncel *et al.* (*Scientific Reports*, Vol. 10, April 2020).

46   'Origin of clothing lice indicates early clothing use by anatomically modern humans in Africa'
      by Melissa A. Toups, Andrew Kitchen, *et al.* (*Molecular Biology and Evolution*, Vol. 28, pp.
      29–32, January 2011).

47   'Neandertals on the beach: use of marine resources at Grotta dei Moscerini (Latium, Italy)' by
      Paola Villa, Sylvain Soriano *et al.* (*PLoS ONE*, https://doi.org/10.1371/journal.pone.0226690,
      January 2020).

48   'Last interglacial Iberian Neandertals as fisher-hunter-gatherers' by J. Zilhão, D. E. Angelucci *et
      al.* (*Science*, Vol. 367, Issue 6485, March 2020).

49   'The evolution and changing ecology of the African hominid oral microbiome' by James A.
      Fellows Yates, Irina M. Velsko *et al.* (*PNAS*, Vol. 118, No. 20, May 2021).

50   'Human oral microbiome cannot predict Pleistocene starch dietary level, and dietary glucose
      consumption is not essential for brain growth' by Miki Ben-Dor, Raphael Sirtoli and Ran Barkai
      (*PNAS*, Vol. 118, September 2021).

51   'Reply to Ben-Dor *et al.*: Oral bacteria of Neanderthals and modern humans exhibit evidence
      of starch adaptation' by Christina Warinner, Irina M. Velsko and James A. Fellows Yates (*PNAS*,
      Vol. 118, No. 37, September 2021).

52   'Oldest cave art found in Sulawesi' by Adam Brumm, Adhi Agus Oktaviana *et al.* (*Science
      Advances*, Vol. 7, Issue 3, January 2021).

53   'The Châtelperronian Neanderthals of Cova Foradada (Calafell, Spain) used imperial eagle
      phalanges for symbolic purposes' by Antonio Rodríguez Hidalgo, Juan Ignacio Morales, *et al.*
      (*Science Advances*, Vol. 5, Issue 11, November 2019).

54   'Pluridisciplinary evidence for burial for the La Ferrassie 8 Neandertal child' by Antoine
      Balzeau, Alain Turq *et al.* (*Scientific Reports*, Vol. 10, December 2020).

55 'Denisovan DNA in Late Pleistocene sediments from Baishiya Karst Cave on the Tibetan Plateau' by Dongju Zhang, Huan Xia *et al.* (*Science*, Vol. 370, pp. 584–87, October 2020).

56 'A late Middle Pleistocene Denisovan mandible from the Tibetan Plateau' by Fahu Chen, Frido Welker *et al.* (*Nature*, Vol. 569, pp. 409–12, May 2019).

57 'Late Middle Pleistocene Harbin cranium represents a new *Homo* species' by Qiang Ji, Wensheng Wu *et al.* (*The Innovation*, Vol. 2, Issue 3, August 2021).

10 THE FIRST SAPIENS

58 'Age of the oldest known *Homo sapiens* from eastern Africa' by Céline M. Vidal, Christine S. Lane *et al.* (*Nature*, Vol. 601, pp. 579–83, January 2022).

59 'Deciphering African late middle Pleistocene hominin diversity and the origin of our species' by Aurélien Mounier and Marta Mirazón Lahr (*Nature Communications*, Vol. 10, September 2019).

60 'Apidima Cave fossils provide earliest evidence of *Homo sapiens* in Eurasia' by Katerina Harvati, Carolin Röding *et al.* (*Nature*, Vol. 571, pp. 500–504, July 2019).

61 'The earliest modern humans outside Africa' by Israel Hershkovitz, Gerhard W. Weber *et al.* (*Science*, Vol. 359, Issue 6374, pp. 456–59, January 2018).

62 'Mitochondrial DNA and human evolution' by Rebecca L. Cann, Mark Stoneking and Allan C. Wilson (*Nature*, Vol. 325, pp. 31–36, January 1987).

63 'An African origin for the intimate association between humans and *Helicobacter pylori*' by Bodo Linz, François Balloux *et al.* (*Nature*, Vol. 445, pp. 915–18, February 2007).

64 'Recovering signals of ghost archaic introgression in African populations' by Arun Durvasula and Sriram Sankararaman (*Science Advances*, Vol. 6, Issue 7, February 2020).

11 JEWELS, BURIALS AND CEREMONY

65 'Early Middle Palaeolithic culture in India around 385–172 ka reframes Out of Africa models' by Kumar Akhilesh, Shanti Pappu *et al.* (*Nature*, Vol. 554, pp. 97–101, February 2018).

66 'Conservation of brain connectivity and wiring across the mammalian class' by Yaniv Assaf, Arieli Bouznach *et al.* (*Nature Neuroscience*, Vol. 23, pp. 805–808, June 2020).

67 'Evolutionary expansion of connectivity between multimodal association areas in the human brain compared with chimpanzees' by Dirk Jan Ardesch, Lianne H. Scholtens *et al.* (*PNAS*, Vol. 116, No. 14, pp. 7101–06, March 2019).

68 'Human TKTL1 implies greater neurogenesis in frontal neocortex of modern humans than Neanderthals' by Anneline Pinson, Lei Xing *et al.* (*Science*, Vol. 377, Issue 6611, September 2022).

69 'A 100,000-Year-Old Ochre-Processing Workshop at Blombos Cave, South Africa' by Christopher Henshilwood, Francesco d'Errico, *et al.* (*Science*, Vol. 334, Issue 6053, pp. 219–222, October 2011).

70 '78,000-year-old record of Middle and Later Stone Age innovation in an East African tropical forest' by Ceri Shipton, Patrick Roberts *et al.* (*Nature Communications*, Vol. 9, May 2018).

12 THE BIG QUESTIONS

71 'Climate effects on archaic human habitats and species successions' by Axel Timmermann, Kyung-Sook Yun *et al.* (*Nature*, Vol. 604, pp. 495–501, April 2022).

72 'Recent acceleration of human adaptive evolution' by John Hawks, Eric T. Wang *et al.* (*PNAS*, Vol. 104, No. 52, December 2007).

73 'The Contribution of Neanderthals to Phenotypic Variation in Modern Humans' by Michael Dannemann and Janet Kelso (*The American Journal of Human Genetics*, Vol. 101, pp. 578–89, October 2017).

74 'Identification of DNA methylation changes in newborns related to maternal smoking during pregnancy' by Christina A. Markunas, Zongli Xu *et al.* (*Environmental Health Perspectives*, Vol. 122, No. 10, October 2014).

75 'Cliff-edge model of obstetric selection in humans' by Philipp Mitteroecker, Simon M. Huttegger *et al.* (*PNAS*, Vol. 113, No. 51, December 2016).

ACKNOWLEDGEMENTS

Writing books takes a squad – science books even more so. Thanks to: my editors Jo Stanstall and Becca Wright for their guidance and patience; my eagle-eyed copy editor Helen Cumberbatch; my meticulous proofreader Monica Hope; my expert readers for their thoughts and suggestions (any inaccuracies are solely mine); Jenny Wild, Cass Parker, Jayne Batzofin, and Helen Scales; PdW, without whose support none of this would have been possible; and, finally, KdW, for teaching me how magnificent humans really are.

INDEX